Spin Management
and
Recovery

Spin Management and Recovery

Michael C. Love

McGraw-Hill

New York   San Francisco   Washington D.C.   Auckland   Bogotá   Caracas   Lisbon
London Madrid   Mexico City   Milan   Montreal   New Dehli   San Juan   Singapore
Sydney   Tokyo   Toronto

Library of Congress Cataloging-in-Publication Data
Love, Michael C. (Michael Charles)
    Spin management and recovery / Michael C. Love
      p.   cm.
    Includes bibliographical references and index.
    ISBN 0-07-038809-1 (hc.). -- ISBN 0-07-038810-5
    1. Spin (Aerodynamics) 2. Emergency maneuvers (Aeronautics)
  I. Title
  TL574.S6L67   1996
  629.132--dc20   96-38271
                                        CIP

McGraw-Hill

A Division of The **McGraw·Hill** *Companies*

1 2 3 4 5 6 7 8 9 0 DOC/DOC 9 0 1 0 9 8 7 6 (PBK)
1 2 3 4 5 6 7 8 9 0 DOC/DOC 9 0 1 0 9 8 7 6 (HC)

ISBN 0-07-038809-1 (HC)
ISBN 0-07-038810-5 (PBK)

The sponsoring editor for this book was Shelley Chevalier, the editing supervisor was Sheila Gillams, and the production supervisor was Suzanne W. B. Rapcavage. This book was set in 10 point Palatino by Benchmark Productions, Inc., Boston, MA.

Printed and bound by R. R. Donnelley & Sons Company.
This book is printed on acid-free paper.

LIMITS OF LIABILITY AND DISCLAIMER OF WARRANTY
The authors and publisher have exercised care in preparing this book and the programs contained in it. They make no representation, however, that the programs are error-free or suitable for every application to which the reader may attempt to apply them. The authors and publisher make no warranty of any kind, expressed or implied, including the warranties of merchantability or fitness for a particular purpose, with regard to these programs or the documentation or theory contained in this book, all of which are provided "as is." The authors and publisher shall not be liable for damages in amounts greater than the purchase price of this book, or in any event for incidental or consequential damages in connection with, or arising out of the furnishing, performance, or use of these programs or the associated description or discussions.

    Readers should test any program on their own systems and compare results with those presented in this book. They should then construct their own test programs to verify that they fully understand the requisite calling conventions and data formats for each of the programs. Then they should test the specific application thoroughly.

To Dottie, Ryan, and Emily, for understanding my needs when it comes to flying, and for freely allowing me the time it takes to do it right.

Contents

CHAPTER 3—
INTRODUCTION TO SPINS   71

Acknowledgments

THANKS TO MY INSTRUCTORS OVER THE YEARS, FOR LAYING THE foundation that has allowed me to continue to learn about stalls and spins. Their professionalism has stood me well over the years. Thanks also to Lee Johnson for proofreading this book and providing feedback on it. A very large thanks to both Claude D. Carlson and Lee Johnson for use of the Pitts S-2Bs used throughout this book. Thanks to Gary Nelson for flying with me and showing me how to do the photography that is such a large part of this book. I would also like to thank Frank Gattolin of the NTSB office in Chicago, IL for pointing me in the right direction to find the NTSB accident reports used in the book. Thanks also to Ray Whipple at NASA Langley Research Center for providing me with information related to NASA spin studies. Finally, thanks to the management and staff at Wisconsin Aviation, Madison, Wisconsin, for the support they have demonstrated over the years.

Introduction

I HAVE BEEN FORTUNATE DURING MY FLYING CAREER, HAVING HAD the good fortune of being taught by competent, knowledgeable flight instructors. These instructors taught me the importance of proper flight training, and how it increases the safety of pilots that take the lessons to heart. The training I received early in my flying included stall and spin entry and recovery. I confess that I have always been fascinated by aerobatics, and these unusual attitudes were my first exposure to the world of flying at the edges of a plane's flight regime.

Because I enjoy them, and knowledge of these maneuvers helps develop safer pilots, over the last twenty-plus years of flying I have tried to learn all I could about stalls and spins. I believe every pilot should be proficient at recognizing the onset of a stall or spin and know the proper techniques for recovering from them. Though spin training is not required by the Federal Aviation Administration (FAA) except for flight instructors, in addition to stall training, I have always taught my students spin entry and recovery. They may never enter another spin after the training is completed; but they will at least understand what a spin is, and more importantly, how to get out of it.

This book reflects my desire to make information related to stalls and spins available to as many pilots as possible. While not intended as a self-taught course in stalls and spins, it will provide valuable information, both technical and practical, to those who read it. Combined with a qualified flight instructor and the proper aircraft, practicing the stall and spin maneuvers outlined in this book will provide you with experience that will make you a safer pilot. It must be emphasized that the stalls and spins covered in the book should first be flown with a qualified flight instructor, and only in a properly certified airplane. Many aircraft today are not certified for spins, or certain types of stalls, and should not be used to practice those maneuvers. The advanced spins that will be covered in later chapters should be done only in an aerobatic aircraft certified for those maneuvers. Be sure to comply with all FAA regulations related to spins, as well.

This book discusses many aspects related to stalls and spins, including many unusual situations. It fits well into a flight school curriculum, giving instructors a teaching aid to help prepare students for stall and spin training. Outside of a flight school curriculum, the book can supplement the training in stalls and spins that pilots have already had.

How to recognize the onset of a stall and how to avoid stalls through proper diligence are among the topics reviewed. Most stall/spin accidents take place during takeoff and landing, with the pilot not even recognizing a stall/spin situation exists until it is too late. Through practicing these maneuvers at safe altitudes, the pilot learns how the controls feel, how the engine sounds, and how to recognize other visual and audio cues that precede a stall. This book provides you with a strong understanding of what you can expect to experience during these situations.

The importance of control coordination is also discussed at length. I have given spin training to pilots who have surprised themselves by entering tight, skidding turns from base to final approach. As they tried to keep the wings from banking too steeply, they used excessive rudder trying to force the nose of the plane around. Predictably, a plane in this situation eventually stalls and breaks, the circumstances worsened by the uncoordinated use of controls. Once pilots have come this close to the edge, they are motivated to take proper training to avoid the same mistakes in the future. As you read, you will learn the importance of the correct use of controls and the rationale behind their use.

The book includes various examples of stall and spins situations, some of which the average pilot may never encounter, but should be able to recognize. In the case of stalls and spins, knowledge is the key to avoidance and recovery.

Chapter 1 defines stalls, discussing facts related to the critical angle of attack, lift enhancement devices, and stall warning characteristics. Slow flight is also covered, providing information on control effectiveness, left-turning tendencies, aircraft feel during slow flight, and entry into and exit from slow flight.

A number of stall scenarios are reviewed in Chapter 2, including normal stalls and accelerated stalls. Also discussed are approach and departure stalls. Entry into these stalls from slipping and skidding turns are reviewed, giving the reader an appreciation of the need for coordinated use of controls. Finally, altitude loss during a stall, and how to reduce it to a minimum, are covered.

Chapter 3 introduces spins. Discussions of aircraft spin certification and Federal Aviation Regulations (FARs) begin the chapter. Spins are then defined and related to stalls. Another section covers how to enter normal spins and recover from them. Positioning of the rudder, elevators, and ailerons during entry into the spin, during the spin, and during recovery are also reviewed. A comparison of spins and spirals is made, including the differences in recovery from each of them. The different phases of a spin and improper spin recovery techniques are also covered in detail. Altitude loss during spins, instrument behavior, and emergency spin recovery techniques round out the chapter.

Normal spins are covered in Chapter 4. An extensive review of entry into normal spins begins the chapter. Use of rudder, elevator, and ailerons during spin entry are covered, in addition to improper spin entry techniques. Spin recovery is discussed next. Standard spin recovery techniques are the first topic of interest, followed by a look at the disorientation that is common among pilots who are not experienced with spins. A review of improper recovery steps, and how they can affect the spin, are also covered. The last major topic is an analysis of aircraft characteristics during a spin, such as how an aircraft's rate of rotation and pitch can change during the spin.

Chapter 5 introduces accelerated and crossover spins. Accelerated spins are the result of improper control use during a spin and result in a rapid increase in the rate of rotation. An unwary pilot often panics when faced with an accelerated spin and compounds the situation through the use of incorrect control inputs. How accelerated spins are entered and how to recover from them are topics reviewed in the chapter. Accelerated spin characteristics are also discussed at length. The crossover spin takes place when improper spin recovery procedures are applied, and an upright spin suddenly becomes an inverted spin. Often the pilot does not even realize the plane has crossed over from an upright to an inverted spin and has no idea why normal spin recovery control placement is not working. By understanding how the crossover spin is caused, pilots can use the proper recovery procedures to avoid them.

Chapter 6 introduces the reader to inverted spins. The chapter begins with an explanation of inverted spins, examining the characteristics and definition of an inverted spin. Next a review of inverted flight, coupled with aircraft requirements, highlights the importance of performing this maneuver in the proper aircraft. Spin entry and spin recovery are reviewed. Inverted spins can be especially disorienting, and a discussion of this topic is included in spin recovery techniques.

Flat spins are discussed in Chapter 7. Once again, this is a spin most pilots never need to contend with, outside of the aerobatic arena. But every year, for a variety of reasons, pilots unexpectedly find themselves in this type of spin. Improper use of controls during spin recovery, aft center of gravity, and other factors can force a plane into a flat spin. This chapter reviews what can cause a flat spin and the best techniques for recovery from it. Both upright and inverted flat spins are reviewed. Especially important is the discussion of potential disorientation during these spins. I have taught flat spin entry and recovery to pilots who have never experienced them before, and the first comments are always how they are extremely disoriented during the spin, having no idea of the aircraft's attitude during the spin. Knowing where to look to maintain orientation is important during any unusual attitude maneuver.

Chapter 8 discusses NASA stall/spin studies and covers a number of stall/spin accidents from reports gathered by the National Transportation and Safety Board (NTSB). NASA has invested a great deal of time and effort in an attempt to better understand how spins affect general aviation aircraft. This information should further your understanding of the mechanics behind spins. As a result, you will be more aware of real-world situations, and how easily unknowing pilots can get into serious trouble. The intent is to re-enforce the need for proper spin training. The NTSB stall/spin accident reports help highlight the point that complacency related to stalls and spins can turn into serious problems.

A number of factors related to spin training are covered in Chapter 9. The chapter begins with topics such as why pilots should receive spin training even though the FAA does not require such training for most pilot certificates. Reasons include pilot safety, an increase in a pilot's skills and knowledge, and additional confidence. For those who do decide to get training, the chapter also discusses what to look for in spin training and the schools that provide it. The school's reputation, the type of aircraft and safety equipment used, the curriculum, and the instructor are all factors that will determine if you come away from the training with useful skills and knowledge, or feel that you learned nothing. Factors affecting pilot judgment round out the chapter. Often we are unaware of the reasons behind our behavior in the cockpit, and a number of behavioral patterns that can lead to poor decisions are covered.

Chapter 10 is intended for flight instructors, but it will benefit any reader. This chapter discusses techniques related to teaching stalls and spins. Preparation of the flight instructor and student prior to the flight can increase the understanding and comfort of the student during the flight portion of the lesson. Topics reviewed in this chapter include aircraft familiarity and lesson preparation for the instructor. Concepts that should be covered during the preflight discussion are also mentioned. These include aircraft attitude during the maneuvers, control inputs, physical reactions and sensations that can be expected, flight safety, and reference points the student should use during each maneuver. Finally, stall and spin lessons are discussed, including how the instructor can show the student the maneuver the first time it is executed, then have the student execute the maneuver.

The conclusion summarizes many of the underlying principles discussed throughout the book. Pilot safety and the proper mind set that should be maintained whenever a pilot flies are also discussed. Please keep in mind that safety should always be the number one concern of any pilot, and that is the motivation for this book. The more pilots know about how to deal properly with each stall and spin situation, the better prepared they are not only to recover from them, but to avoid them. Every year pilots end up as stall/spin accident statistics due to a lack of knowledge. Please take the time to get the proper training and use the information in this book to enhance your safety as much as possible.

1

Introduction to Stalls

THERE ARE A NUMBER OF REASONS FOR WRITING THIS BOOK. FIRST
of all, I enjoy executing stalls and spins, as well as teaching them to my students. Second, by publishing information related to stalls and spins, I can reach
a larger audience. Those who have questions related to stalls and spins, or who
have not been exposed to stalls and spins during flight training, should find
answers as they read the book. For many pilots, stalls and spins represent an
unknown, and feared, flight regime. Many of my students express an initial
concern related to both stalls and spins, indicating that they received only a cursory introduction to the topics during previous flight training. It is my firm
belief that all pilots should be familiar with stalls and spins, even if they never
plan on practicing the maneuvers outside of biennial flight checks. Finally, I am
motivated by the desire to help pilots achieve the highest level of safety possible when they fly. By understanding what factors are involved in entering stalls
and spins, you can more easily prevent inadvertent entry. As a result, flight
safety increases and you, as the pilot, are more comfortable with the plane's
flight envelope.

Once again, this book is not intended as a self-taught guide to stalls or
spins. It provides additional information to supplement a flight training course
or to be used as a refresher on stalls and spins. Always have a flight instructor
teach you the proper techniques during dual instruction and before performing
any maneuver described in this book, and be certain the plane you are flying is
certified for each maneuver. Today, many general aviation aircraft are placarded against certain types of stalls and spins. There is a reason for this, related
to aircraft certification procedures, which will be covered in detail in Chapter 2.
Finally, be certain to adhere to all FAA regulations for visibility, cloud clearances, and other regulations when you execute any stall or spin.

This chapter reviews a number of topics, beginning with the definition of
stalls. Other topics include slow flight, or flight at minimum controllable

airspeed (MCA), normal stalls, accelerated stalls, approach stalls, and departure stalls.

DEFINING STALLS

A number of factors determine what causes a plane to stall. Understanding how lift is generated, and how it is lost during a stall, is a significant issue. Each plane has different stall characteristics, but what causes loss of lift is consistent from plane to plane. Understanding how lift is created by the wing, and how lift is affected by the angle of attack, can help you understand not only how to enter stalls, but also how to avoid them.

DEFINING LIFT

Lift is created as a result of two forces, pressure differential and deflection pressure. The pressure differential portion of lift generation is due to the shape of the airfoil used in an airplane's wing design. Air traveling over the upper surface of the wing travels a longer distance than air flowing along the underside of the wing. According to the laws of physics, the air along the upper surface must travel faster than the air over the lower surface, resulting in a low pressure area forming along the upper wing (Figure 1-1a). As the speed differential increases, so does the difference in pressure between upper and lower wing surfaces. The higher pressure on the lower wing surface "lifts" the wing (Figure 1-1b), providing a portion of the lift necessary for the plane to fly.

The second force creating lift is deflection lift. As air flows along the underside of a wing, it is deflected downward. An equal and opposite reaction takes place and the deflected airflow pushes upward against the wing and plane (Figure 1-1c), also causing the plane to fly. Both types of lift, pressure and deflection, increase as the speed of the wing moving through the air increases. Therefore, the faster the plane is flying, the greater the lift generation capability of the wing. Pressure and deflection lift also increase as the angle of attack increases. A greater angle of attack produces greater lift, up to a point. Once a maximum angle of attack has been reached, the flow of air over the wing's surface becomes so turbulent that lift is reduced.

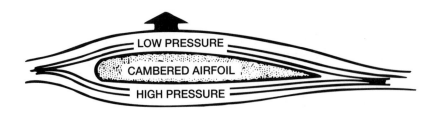

Fig.1-1a Airflow around wing.

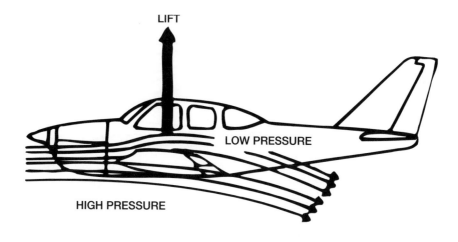

LIFT

LOW PRESSURE

HIGH PRESSURE

Fig. 1-1b Pressure differential lift.

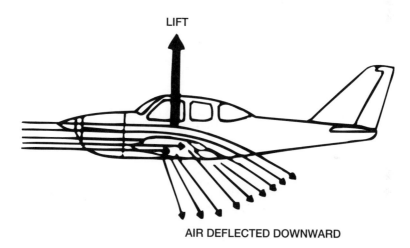

LIFT

AIR DEFLECTED DOWNWARD

Fig. 1-1c Deflection lift.

ANGLE OF ATTACK

When an aircraft is moving through the air at a relatively slow airspeed, such as during takeoff or landing, a higher angle of attack is essential to achieve the lift from pressure and deflection that is required to keep the plane flying. The angle of attack is defined as the angle formed between the relative wind and the wing cord line. Figure 1-2 illustrates a wing at three different angles of attack. At higher relative speeds, the wing is able to generate sufficient lift at a low angle of attack. As the plane slows, a medium angle of attack is required to maintain the necessary lift. Finally, as the plane slows further still to an insufficient airspeed, the angle of attack becomes so large that air is no longer able to

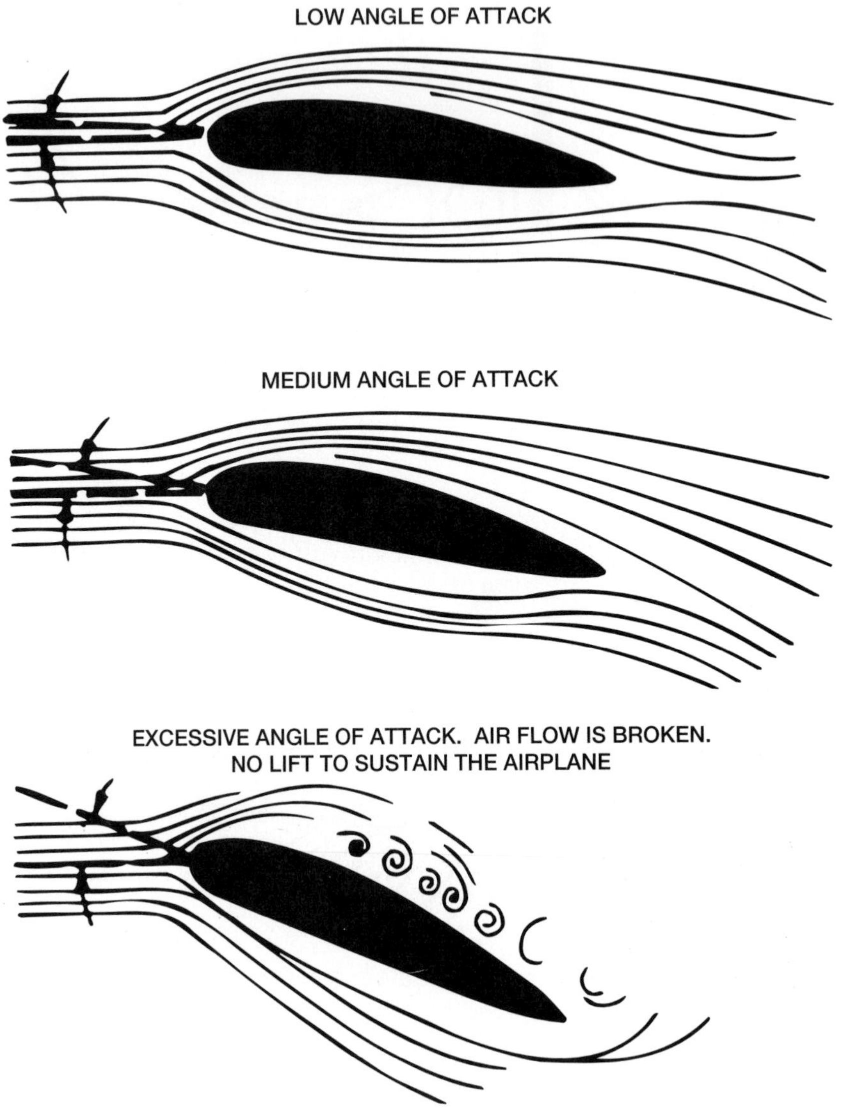

LOW ANGLE OF ATTACK

MEDIUM ANGLE OF ATTACK

EXCESSIVE ANGLE OF ATTACK. AIR FLOW IS BROKEN.
NO LIFT TO SUSTAIN THE AIRPLANE

Fig. 1-2 Low/medium/high increasing angle of attack.

maintain a smooth pattern of flow over the wing's upper surface and becomes turbulent. This turbulence causes an increase in pressure along the top of the wing, reducing the pressure differential between the upper and lower surfaces of the wing. This reduction in the pressure differential also results in a reduction in lift.

Critical Angle of Attack

When airflow over the upper surface of the wing becomes so turbulent that the wing can no longer generate sufficient lift, the wing has reached the critical angle of attack, or the point at which the wing stalls. While this angle varies somewhat from plane to plane, it generally runs between 15 and 20°.

Note that the critical angle of attack can be reached even when the plane is in level flight, developing full power from the engine. This is a significant point to remember. The critical angle of attack will be achieved when the angle between the relative wind and the cord line of the wing exceeds 15 to 20°. Figure 1-3 illustrates how the flow of air around the wing changes as the angle of attack increases from 4 to 20°. Note how the airflow becomes turbulent toward the back of the wing, as it progressively moves forward. Wings are designed to have certain stall characteristics, and the shape of the airfoil and wing planform determine how the stall progresses over the wing.

Figure 1-4 shows a number of different wing planforms, or the shapes of the wing as seen from above, and the stall progression pattern for each of them. Many aircraft today are designed to have the wing root stall first, with the outboard portion of the wing still producing lift. This helps maintain aileron authority during a stall and increases your ability to control the attitude of the plane while it is stalled.

Flying an airplane at slow airspeeds and high angles of attack can result in the critical angle of attack being exceeded. However, abruptly pulling back on the control yoke at cruise flight can have the same effect, resulting in a high-speed stall. You should also be aware that high-speed stalls can be not only damaging to your airplane's structure, but can cause a violent reaction from the plane. In this situation the plane's nose may not rise a great deal, but the sudden high g loading can cause the plane to stall as readily as during slow flight. Stalls are defined in detail in Chapter 2.

Severe turbulence can also cause a plane to stall as rapid changes in relative wind cause the critical angle of attack to be exceeded. To help avoid potential damage to an aircraft's structure, aircraft manufacturers determine the

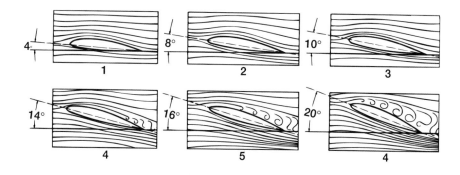

Fig. 1-3 Increasing angle of attack series.

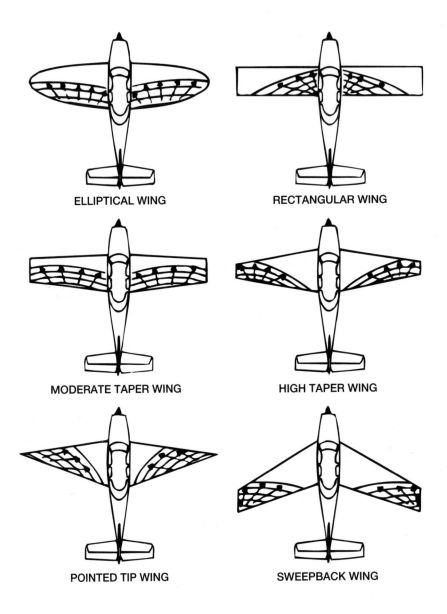

ELLIPTICAL WING

RECTANGULAR WING

MODERATE TAPER WING

HIGH TAPER WING

POINTED TIP WING

SWEEPBACK WING

Fig. 1-4 Wing planforms.

maneuvering speed for each plane. The maneuvering speed is the airspeed at
or below which full, rapid control inputs can be made without causing damage
to the plane. The plane will stall before loads on the wings and airframe result
in damage. For this reason, you should reduce your airspeed to at or below
maneuvering speed when you encounter turbulent air. Flying at speeds above
maneuvering speed causes the plane to stall at higher relative loadings, placing

higher stresses on aircraft structures that could damage or destroy the plane. Be aware that the maneuvering speed changes as the weight of the aircraft changes. Reference your aircraft's operations manual for specifics regarding the correct maneuvering speed for a given weight.

LIFT ENHANCEMENT DEVICES

Early aircraft were basic in their design, the wing lacking lift enhancement devices, such as flaps, that are common on today's planes. As aircraft engineers looked for ways to reduce the stalling speed of an aircraft (and as a result the takeoff and landing speeds), they came up with a number of modifications to the design of the wing. They found that by increasing the camber, or curvature of the airfoil, they could lower the stall speed of the aircraft. Figure 1-5 shows a number of different airfoil designs, each with different airfoil cross-sections. The early airfoil had a large curvature, or camber, on the underside of the wing. While this provided slower stalling speeds, it also resulted in slower cruise speeds. The later airfoil was a compromise, with less camber on the underside of the wing. Finally, the Clark "Y" airfoil became a popular design for general aviation aircraft. While each provided higher cruising speeds for planes, as the camber changed, stall speeds of aircraft increased.

Aeronautical engineers designed lift enhancement devices that allowed them to take advantage of the higher airspeeds that the new airfoil designs allowed, yet helped reduce the stall speed for a plane. These devices changed the camber of the airfoil temporarily, such as during takeoff and landing. Some also increased the ability of the air to flow smoothly over the upper surface of the wing at high angles of attack. Flaps, slots, and slats help airplanes fly at slow airspeeds during takeoff and landing by changing the shape of the airfoil and reducing the stall speed. Understanding how each of them affect the stall speed will be useful as you fly planes equipped with these devices.

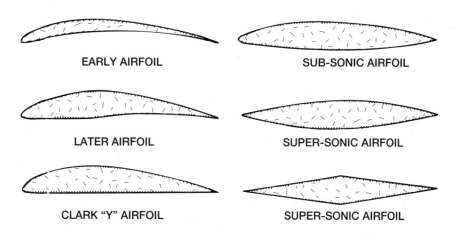

EARLY AIRFOIL	SUB-SONIC AIRFOIL
LATER AIRFOIL	SUPER-SONIC AIRFOIL
CLARK "Y" AIRFOIL	SUPER-SONIC AIRFOIL

Fig. 1-5 Airfoil shapes.

Flaps

There are many different types of flap designs. Figure 1-6 illustrates three common flap configurations: the plain flap, the split flap, and the fowler flap. As can be seen, each of these flap implementations results in a greater curvature of the airfoil when the flaps are in the extended position. In the retracted position, they return the airfoil to its original design cross-section.

The first flap design, the plain flap, hinges at the trailing edge of the wing. When lowered, the flap pivots down around its hinge points. Flaps of this design often allow the pilot to position the flaps at predetermined increments through either mechanical or electrical actuators. These increments vary from one plane design to another, but frequently can be set at 10, 20, 30, and 40° of

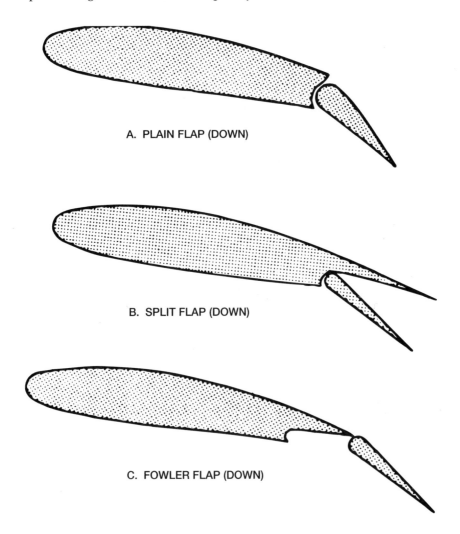

A. PLAIN FLAP (DOWN)

B. SPLIT FLAP (DOWN)

C. FOWLER FLAP (DOWN)

Fig. 1-6 Plain, split, and fowler flap types.

flap extension. Other aircraft allow only three flap setting positions to be selected, but still provide adequate ranges up to 40° of flap extension.

In addition to providing additional lift capability, flaps generate additional drag as they are extended into the wind stream. This increase in drag lets the plane descend using a steeper glide slope without increasing the airspeed beyond safe limits. This capability is useful for various types of landings, such as in short or soft field situations.

The second design, the split flap, also changes the camber of the airfoil, but pivots at a point under the wing. As illustrated, the upper surface of the wing retains its original shape. This type of flap design seems popular on certain twin engine aircraft, possibly due to the location of engine nacelles and the cost associated in designing flaps around these structures.

The fowler flap is the last example. In addition to moving down, fowler flaps also move back. This not only changes the camber of the wing, but increases the surface area of the wing. Various Cessna single engine aircraft have used this design, providing them with excellent takeoff and landing characteristics. Fowler flaps are also popular on many airliners, with three to four separate sections comprising each flap mechanism. This design allows significant reduction in the speeds that are necessary to land these large aircraft; but, when the flaps are retracted, the plane can cruise at Mach numbers in excess of .8.

Flaps have a significant impact on the stall characteristics of some general aviation aircraft, and seemingly little effect on others. On some aircraft, stall speeds are reduced by ten to fifteen knots; while on others, flaps may produce only a few knots of reduction. Anyone who has witnessed the short field takeoff and landing capabilities of STOL airplanes can attest to the fact that these planes seem to be able to fly at little more than walking speed. The size of the flaps, their design, and the design of the airfoil all affect the effectiveness of the flaps. Remember to use flaps in accordance with the aircraft's operations manual.

Slots

Leading edge slots were popular on aircraft built several decades ago. The first airplane equipped with slots to which I was exposed was a Stinson taildragger. A number of other aircraft designs also included slots in the forward portion of the wing, such as the Globe Swift. Figure 1-7 provides a cross-section view of wings with and without leading slots. In this example, both wings are at the same angle of attack. The wing with the slot allows air to flow through the slot, in addition to over the leading edge of the wing. This has the effect of allowing a smoother flow of air over the upper surface of the wing as compared to the wing without the slot. This smoother airflow lets the wing fly at a slower airspeed before stalling.

While not in general use on newer aircraft, leading edge slots can be found on some short field takeoff and landing aircraft that are being manufactured today. For special-application aircraft that are designed specifically for

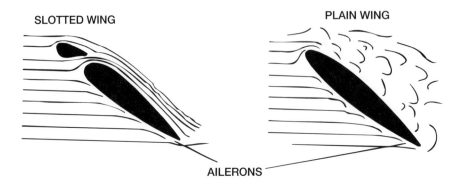

SLOTTED WING   PLAIN WING

AILERONS

Fig. 1-7 Slotted wing leading edge.

slow-speed environments, slots help the aircraft designer to achieve low stalling speeds. Once again, if you fly a plane equipped with slots, you should read the aircraft's operations manual to understand their effect on the stalling speed of the plane.

Other Lift Enhancement Devices

Aircraft have been equipped with other types of devices to reduce stalling speed in addition to flaps and slots. While not common on general aviation aircraft, leading edge flaps are prevalent on some airliners. These generally pivot forward and down from the leading edge of the wing, having the effect of increasing the camber of the underside of the airliner's airfoil. Figure 1-8 demonstrates how leading edge flaps can work. When combined with fowler flaps, a dramatic change in the shape of the airfoil can be achieved.

Leading edge slats are also installed on some aircraft. Like leading edge flaps, these are not common on general aviation aircraft. Leading edge slats are designed to allow the leading edge of the wing to move forward and down, usually on guides and rollers. Figure 1-9 is an example of leading edge slats. These devices cause a change in the shape of the airfoil and help reduce the stall speed of the plane. On some designs when the slats are in the extended position, they also open a slot in the wing, combining the change in airfoil shape with the positive effect of a slotted wing.

The next time you fly on an airliner, notice the devices on the leading and trailing edges of the wing. The entire shape of the wing changes as the flight crew prepares the plane for takeoff while they taxi out to the runway. Then, once the plane is at safe flying speeds, these devices are retracted. This design allows the plane to land at safe airspeeds and still maintain the cruise speeds that are required.

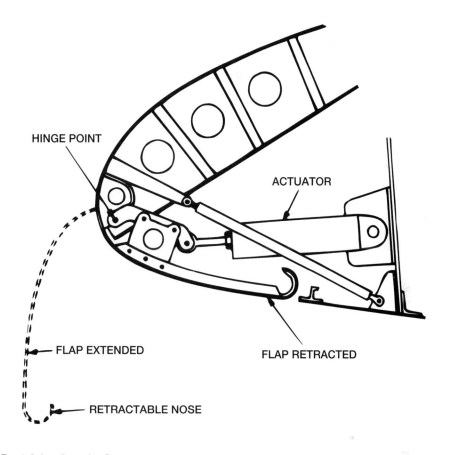

HINGE POINT

ACTUATOR

FLAP EXTENDED

FLAP RETRACTED

RETRACTABLE NOSE

Fig. 1-8 Leading edge flaps.

STALL WARNING CHARACTERISTICS

There are a number of ways that you can tell if a plane is approaching stall speed. To pilots who are in tune with a plane, these signs are unmistakable. When you fly, you should be aware of these signals; and at their onset, you should immediately take corrective action to avoid a stall. Each plane you fly will have different stall characteristics, but the principles discussed here apply to all aircraft to one degree or another. Be sure to review your plane's operations manual for information specific to the plane you fly. You may find the manufacturer has additional information to offer in stall warning and recovery that will help you avoid a stall, or minimize altitude loss during recovery from a stall.

This section discusses several areas related to stall warning, including hearing, control feel, aircraft motion (kinesthesia), stall buffet, and aircraft attitude (vision). Under normal stall conditions, one or all of these indicators will give you warning that the plane is approaching stall speed.

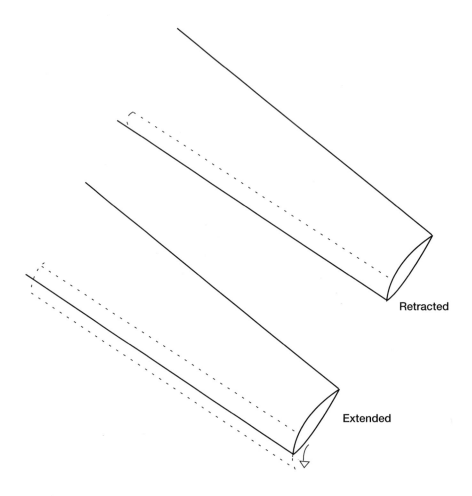

Fig. 1-9 Leading edge slats.

The sound your airplane makes is one way to tell the plane may be approaching stall speed. As a plane slows, the sound of air flowing over it changes, often becoming quieter. The engine sound may also change as the load on the engine changes. In the case of fixed pitch propellers, the RPM may change. As you are flying the plane, you should listen to it. Whether you are in climb mode after takeoff, executing a landing descent, or at cruise flight, the plane will assume a certain level and set of sounds. Listen to the sounds your plane makes and see how they change as you increase or decrease the speed of the plane. Notice how the plane becomes quieter as it slows. This will be a very good indicator to you that the plane is slowing even if you are not looking outside the plane or at the instruments.

Control feel is another signal that you may be approaching stall speed. As a plane slows, the reduced flow of air over the wings will cause the controls to become "softer." With less pressure against them, the controls will move easier

Retracted

Extended

against the slipstream than at higher airspeeds. The plane will also be less responsive to your control inputs. For example, it requires greater aileron inputs at slower airspeeds to bank the airplane than is necessary at higher speeds. As noted throughout this book, an awareness of aircraft feel can give you a great deal of information about how the aircraft is flying. Banks are the result of lift differential between the wings, the wing with the lowered aileron generating greater lift than the wing with the raised aileron. Just as the wing requires a greater angle of attack to generate lift at slow speeds, it also needs more aileron input to achieve the necessary lift differential to bank the plane. You also need greater elevator inputs at these slow speeds to maintain a given pitch angle.

Aircraft motion is another method of detecting a reduction in airspeed and the possible onset of a stall. As you fly, your body becomes accustomed to the "steady" forward motion of the plane. The body also senses changes in the speed of the plane as it accelerates and decelerates, climbs and descends, or pitches nose up and down. This sensitivity to changes in motion, also known as kinesthesia, allows you to very quickly detect changes in your plane's speed as it moves through the air. If, for one reason or another, you do not notice airspeed dropping off on the airspeed indicator, you should be able to detect the deceleration without looking at the instruments. Not only will the plane slow as it approaches a stall, but the attitude may change and the plane may lose altitude as it "mushes" along. The more familiar you are with the plane you fly, the more adept you will become at noticing these changes in speed.

Stall buffet is another warning sign that you are very close to a stall. As your plane approaches the critical angle of attack, the airflow over the wings becomes turbulent. As the angle of attack continues to increase and the airflow's turbulence continues to increase, it begins to buffet the wing, fuselage, and tail surfaces. This buffeting can range from an almost unnoticeable vibration to a pounding that shakes the entire aircraft. The onset of stall buffet is a sure sign that you are very near to the stall angle of attack. If you recognize that buffeting has set in, you can avoid the stall by reducing the angle of attack and allowing the airflow to become smoother again. You should also be aware that all control surfaces, not just the wing, can be stalled.

On a recent flight with a student in his Piper Comanche, we practiced both power on and off stalls. When we executed power on stalls, the buffet was very pronounced, the entire elevator shaking as rough air from the wings slapped against them. There was no mistaking when the buffeting started; the entire airplane shook. I have also flown planes that were less obvious as stall buffet set in. You need to experiment with the planes that you fly to find out when stall buffet sets in and how pronounced it becomes. It is worth the effort, though! This knowledge can be valuable in determining just how close to a stall you really are.

Even at high airspeeds, if excessive elevator is applied that produces a highs-peed stall, the airplane will likely generate a high-speed buffet prior to the stall. Use caution with this maneuver, though. A high-speed stall, if flown above maneuvering speed, could damage the aircraft's structure. Also called a whip stall, many airplanes are placarded against this maneuver. To become more attuned to the plane as it approaches stalls, find a qualified flight instructor and practice entering and recovering from stalls. With proper training, you will notice the onset of stall buffet even if you are distracted by other events.

Aircraft attitude is another method of determining that you are approaching a stall. While an aircraft can stall at any airspeed or attitude, the most likely situation is during those times you fly at slow airspeed, such as takeoff and landings. During slow airspeed operations, the plane increases its nose up attitude during an approach to stall. In many cases, the higher angle of attack may block your forward view, reducing your ability to see the horizon ahead. In other aircraft, even if the horizon ahead is not blocked, you may find that the "sight picture" forward over the nose changes significantly enough to give you warning that the pitch angle, and the airspeed, are not what they should be.

As you learn these pitch and attitude characteristics about your plane, you will be able to tell merely by looking out the window, how close you are to the correct angle of attack. During landing, assuming that you use constant flap and power settings, the pitch angle should change very little throughout the approach. If you notice that the pitch angle is creeping up during the approach, it may be a sign that you are not maintaining a constant pitch angle, or airspeed. The higher the relative pitch angle, the greater the probability that you are getting too slow. You should practice watching the pitch changes as the plane slows during both departure and approach stalls. Please keep safety in mind as you practice any maneuver, though. If you are out of practice, or uncomfortable with executing the maneuvers, find a flight instructor to give you instruction until you become more proficient.

This section reviewed several different cues your plane will give you as it approaches a stall. The more you practice executing stalls, with the proper supervision, the better you will become at recognizing each of these cues. The first rule in any emergency situation, such as stall recovery, is to keep flying the plane. Too often pilots begin to focus on a single instrument, or other aspect of a problem they are dealing with, and forget to maintain control of the plane. This is often the scenario for unexpected stalls. In any situation, relax, fly the plane, and listen to what it is telling you.

SLOW FLIGHT

Slow flight practice is an excellent means of learning how the planes you fly behave as they approach stall speed. This type of practice lets you become familiar with each of the cues that were discussed in the last section. In addition, slow flight, also called flight at minimum controllable airspeed (MCA), gives you the opportunity to hone your flying skills in areas such as airspeed control, attitude control, and altitude control. The more you practice flight at

MCA, the more polished you will become during both takeoffs and landings, which are the times when stalls are least wanted and most likely to take place.

This section reviews a number of topics related to slow flight. Among them are control effectiveness, the left turning tendency of aircraft at high angles of attack, pitch angles, "feel" during slow flight, as well as entry into and recovery from slow flight. Stall awareness and avoidance are covered for each topic.

Control Effectiveness

Your plane's control surfaces become less effective as the plane slows. Figure 1-10 represents a plane in slow flight at a constant altitude. Notice the high relative angle of attack. Due to the lower airspeed and reduced amount of air flowing across each control surface, it requires a greater movement of the elevators, rudder, and ailerons to make the airplane behave in the desired manner. This situation is often referred to as the controls being "mushy" or the plane "wallowing" as you maintain wings level flight.

You will also notice the lack of control authority when making turns at MCA. The plane may have a tendency to settle downward as you roll into a bank. Yaw may also be more pronounced. As you may recall, the lowered aileron causes more drag, causing the plane to yaw in the direction opposite the turn. Also known as adverse yaw, the rudder is used to overcome this yawing tendency and keep the plane coordinated during the turn. The need for greater aileron movement causes even more adverse yaw and the need for larger rudder inputs to compensate. Reduced control effectiveness also affects the rudder, though, and even more rudder movement will be necessary to keep the turn coordinated.

As you fly at MCA, be certain not to overcontrol the plane through excessive control inputs. It is possible to cause the plane to inadvertently enter a stall or spin through control inputs that are too large. The more you practice slow flight, the more adept you will become at recognizing how much control input should be used in a given situation. Finesse on the controls during flight at MCA will help you increase the smooth feeling the plane can achieve even at speeds just above a stall.

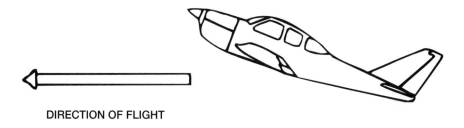

DIRECTION OF FLIGHT

Fig. 1-10 Slow flight at constant altitude.

As you practice flying at MCA, keep in mind that when you execute turns the stall speed will increase. If you are too slow, or bank too steeply, the plane may stall during the turn. You have two basic options to avoid stalling in this situation. The first is to reduce the angle of the bank to the point at which the plane will not stall. The second is to maintain the bank you want, but increase the airspeed. Figure 1-11 is a chart showing how stall speeds increase as an airplane banks. Notice that in a 40° bank the stall speed increases almost 15 percent. If you are flying at 60 miles per hour (mph) in a plane that stalls at 55 mph and enter a 40° bank, the stall speed increases to over 63 mph, quickly resulting in a stall. This information should be of particular interest to you during take-offs and landings, where you are flying at slower airspeeds and may be making turns relatively close to the ground. It becomes easy to see how pilots who may be a little slower than they should be, attempting to make a steep, tight banked turn to final, could find themselves in a stall only a few hundred feet above the ground.

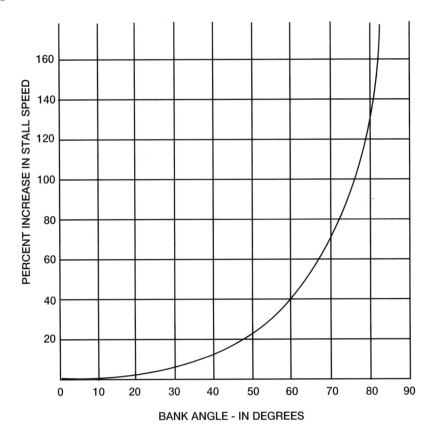

Fig. 1-11 Bank vs. stall speed.

Pilots can avoid these situations by knowing their airplane, including its control feel. A definite warning sign that they are getting critically slow is the loss of control responsiveness. In the above example, a pilot should be able to recognize the problem without looking at the airspeed indicator because the control feel, angle of attack, and sounds of the plane provide ample warning that a stall is approaching. Be sure to practice flight at MCA in the planes you fly to become comfortable with how the plane reacts. If you are out of practice, or uncomfortable with slow flight, take a qualified flight instructor along to help increase your awareness and understanding.

Left-Turning Tendency

This section covers the four forces that you should be aware of when flying a plane at minimum controllable airspeed. Torque reaction, precession, slipstream, and P-factor each affect the plane you fly at all times, but they have the largest impact during slow flight. As a result of these physical forces acting on the plane, it will have a tendency to turn to the left for aircraft with clockwise-turning propellers as seen from the cockpit. If you happen to fly a plane with a counterclockwise-turning propeller, such as a Kitfox homebuilt or Sukhoi aerobatic airplane, these turning tendencies will be to the right. In any case, the principles are the same; you just need to compensate with opposite actions.

Torque reaction is the first of the four factors to review. Sir Isaac Newton stated that "for every action there is an equal and opposite reaction." As the propeller turns in its clockwise direction, it imparts a reaction that causes the plane to roll in a counterclockwise direction, or to the left. Figure 1-12 illustrates this action and reaction concept. As a result of this left-rolling tendency, the plane tries to turn to the left. The more horsepower (HP), or the larger your prop, the greater this tendency is. This reaction will become more pronounced as your plane slows, such as during MCA practice or in a climb. In these conditions, you will be in a nose high attitude, with the engine developing high power, all of which contribute to a stronger reaction to the torque. I have the good fortune of flying Pitts Specials, with 260 HP engines and large, heavy props. As a result of this combination, the plane has a very strong left-turning tendency in slow flight. The light weight of the plane and large engine make it an ideal candidate for torque-related turning tendencies.

Slipstream is the second left-turning tendency. As the propeller turns, it creates a rotating mass of air that flows down the length of the fuselage. Notice slipstream in Figure 1-12, which graphically represents the motion of a slipstream around the fuselage of the plane. As the slipstream flows toward the rear of the plane, it eventually strikes the vertical stabilizer and rudder on their left side, pushing the tail to the right and the nose of the plane to the left. At cruise speeds the slipstream lengthens and its affects are reduced. However, during slow flight, the slipstream is shorter and has a greater impact, resulting in a stronger left-turning tendency. Once again, when the engine is producing a large amount of power, at slow airspeeds, is when the turning tendency is the

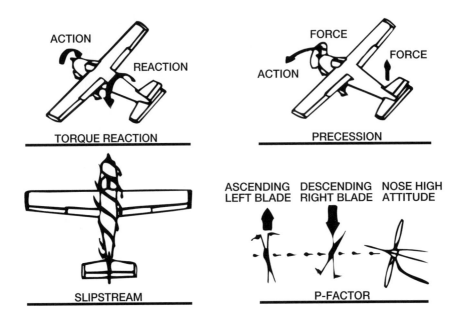

Fig. 1-12 Four left-turning tendencies.

greatest. Aircraft with high horsepower engines will also have a greater inclination to veer to the left than those with smaller engines, due to the increased force of the slipstream.

Precession is the third force that causes a plane to turn. If you have ever tried to move a spinning gyroscope, you have noticed that it resists movement on any of its three axes. This resistance to movement is what made them ideal for use in early aircraft and spacecraft navigation systems. This same tendency manifests itself in a turning propeller on the plane you fly. The propeller resists movement up, down, and sideways, such as when you change the aircraft's attitude or direction of flight. Figure 1-13 depicts how when a force to move the gyroscope is applied, it actually results in movement 90° in the direction of rotation from the point at which the force was applied. In Figure 1-13 you can see that a force applied at the top of the gyroscope causes the gyro to turn to the left, not pitch down as would be expected.

Returning to Figure 1-12, you can see that when the nose is pushed down, the propeller has a tendency to turn the plane to the left. During takeoff, conventional gear planes exhibit a movement to the left as the plane's tail is raised during the takeoff roll. I have noted a definite need to use rudder to counteract this left turning in the tailwheel aircraft I fly. I recently had the opportunity to fly a Kitfox, and due to the counterclockwise-turning engine, I had to use left rudder to overcome the right-turning disposition as the tailwheel was raised from the ground.

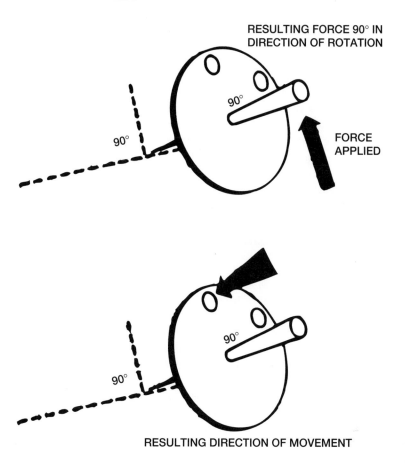

RESULTING FORCE 90° IN
DIRECTION OF ROTATION

90°

90°

FORCE
APPLIED

90°

90°

RESULTING DIRECTION OF MOVEMENT

Fig. 1-13 Gyroscopic precession.

Gyroscopic precession is a force you need to be aware of during slow flight, stalls, and spins as well. As a result of the action of pitching the nose down during recovery from these flight attitudes, you may find that your plane yaws to the left. To keep a proper heading in these situations, you may need to apply opposite rudder. The amount will vary for each aircraft, so you will need to build experience in each plane you fly to get a feel for the correct amount.

The fourth and final force is P-factor. Figure 1-12 compares the difference in the angle of attacks between the ascending left blade and the descending right blade of the propeller when the plane is in a climb. As you can see, the descending blade has a larger angle of attack than the ascending blade. The descending blade generates greater thrust as a result of this difference, and an asymmetrical thrust situation develops. This additional thrust forces the nose of the plane to the left, right rudder then becoming necessary to overcome this yawing. When you do not compensate with right rudder, the ball in the turn and bank indicator will move to the right, indicating that the plane is in an uncoordinated control situation.

At this point you may be asking yourself, "Why doesn't the plane turn to the left when I'm flying at cruise?" The designers and engineers have incorporated several features to overcome the left-turning tendency for the flight regime the plane spends most of its time at, which is cruise. On many aircraft the engine is canted several degrees to the right, in effect helping to turn the plane to the right. The vertical stabilizer may also be offset several degrees from the plane's centerline, once again helping the plane to overcome the forces turning it left. Other design features, such as wingtip washout, can also be used to help the plane fly straight at cruise flight. The left-turning forces are present during all phases of flight, their overall affect on the plane will vary, though.

During flight at MCA, you will notice that the plane drifts off heading during nose high attitudes if you do not compensate for P-factor. The higher the angle of attack, the greater the horsepower and the more thrust that is being developed, the stronger P-factor will be. If you are executing turns while at MCA, you may need right rudder even when turning to the left to make a coordinated turn. Turns to the right require even more right rudder to keep the ball centered. I have seen many students and rated pilots underestimate the amount of rudder required to keep the ball centered in these slow flight situations. Be sure to practice this in the plane you fly so that you become comfortable with the amount of control movement and pressures necessary to maintain coordinated flight. After a short time you will not even need to look at the ball to know that it is centered. The feeling in the seat of your pants will let you know if you are flying in a coordinated manner.

When I was an impressionable sixteen-year-old, my private pilot flight instructor drilled it into my head that the ball should remain centered not only during turns, but during climbs and descents. I was fortunate to have someone so insistent on the importance of coordinated flight. As a flight instructor, I reenforce this principle with my students. A number of students who have come to me for spin training tell me stories about making uncoordinated turns during which the plane suddenly stalls and breaks. This is a graphic example to them that keeping the ball centered is not just something that flight instructors like to harp on, but a way to avoid potentially dangerous stalls or spins close to the ground. I have always felt it is the mark of a pilot that understands their plane when they fly me through a series of stalls, keeping the ball centered in the nose high attitude. Every once in a while, a student will not keep the ball centered, and the plane will begin to enter a spin as it stalls. For this reason alone, it is important that you practice keeping the ball centered at all times to the point that it becomes second nature.

Pitch Angle/Forward Visibility

This section discusses pitch angle and forward visibility while you are flying at MCA and stall airspeeds. Today, many flight schools do not teach complete entry into stalls, but merely an approach to the stall. While I am sure there are reasons for this approach, pilots trained in this manner are not given the opportunity to experience a complete stall. This approach not only leaves a gap in their knowledge, but can leave students wondering why they were not allowed to completely stall an airplane. A certain mystery will shroud a true stall, and this may ultimately become a fear related to an unknown situation. Associated with a complete stall is a nose high angle of attack that can be experienced only when a complete stall is entered. This nose high angle can become even more pronounced when the pilot executes a power on stall. I have had a number of students who have shown discomfort with the angle of attack that is necessary to stall a plane because they did not know what to expect. The fact that many flight instructors do not teach full stalls only reinforces their feelings that a complete stall is a dangerous and uncontrollable flight maneuver.

As with any flight maneuver, if you are not proficient at it, or feel uncomfortable with it, find a qualified flight instructor to fly with you. When you fly at very slow airspeeds, you will notice it becomes difficult, if not impossible, to see forward over the nose of the plane. Figure 1-14 is a picture over the nose of a Cessna 152 from the pilot's seat. As you can see, with power on the nose of the plane almost completely blocks what the pilot can see ahead. Figure 1-15 is a picture down the left wing of the plane, which effectively illustrates the angle of attack as related to the horizon. If pilots keep their eyes forward over the nose of the plane, they will most likely see only blue sky when they are looking forward.

As the plane's angle of attack increases, much of the source of discomfort that students feel is related to an inability to determine the attitude of the plane

Fig. 1-14 Cessna 152 at MCA: Forward view.

Fig. 1-15 Cessna 152 at MCA: Left side view.

correctly by looking in the normal positions outside the plane. To see forward during slow flight, or just before entering a stall, you may need to lean forward to be able to see over the nose of the plane. Looking just to the side of the instrument panel or out the side window of the plane will also help you accurately gauge your plane's attitude. This technique, in addition to the instruments inside the plane, will help you to better orient the position of the plane during the maneuver. Not only will this enable you to determine the plane's pitch, but it also lets you see if the plane is changing direction. How the plane is moving in relation to roads or fence lines can give you a very accurate indication relative to your heading.

The more you practice flying at MCA, the more comfortable you will become with the pitch angles and the use of a scan that helps you to keep them. Many pilots today have fallen into the habit of looking only straight ahead, or inside the plane. During the relatively high pitches found while at MCA, they lose their ability to use visual references outside the plane. Remember not to become a victim to this pitfall, and keep your scan outside the plane, looking to the sides of the plane as you maneuver at slow airspeeds with reduced forward visibility.

Control Feel During Slow Flight

Many pilots rarely or never practice slow flight or stalls once they have completed their private pilot rating. Over time they begin to lose familiarity with the feel of controls during the execution of these maneuvers and only practice flying in these regimes to the extent that the instructor who gives them their biennial requires. This practice every two years may meet the necessary requirements, but it hardly allows a pilot to maintain proficiency. This section briefly reviews the difference in how the controls react and feel during slow flight, takeoff or landing, and normal cruise. Not only understanding that the differences exist, but learning to feel their characteristics, will help make you more aware of what the plane you are flying is doing. This is part of the process of developing a "sense" of what the plane is telling you, without looking at instruments or outside the plane. While flying by not referencing the instruments is not something I recommend on a regular basis, having a feel for the plane can be the "warning flag" that alerts you when you are distracted by something and not paying attention to those instruments.

As mentioned earlier, as the speed of the plane changes, so does the amount of air flowing over the control surfaces and the pressure that this airflow generates against them. At higher airspeeds, such as cruise flight, the air generates a considerable amount of backpressure against each control surface. This back pressure translates itself to you as a solid feel to the control as you exert force to move it via the control yoke or rudder pedals.

At the other end of the spectrum, when you are flying just above stall speed, the plane's controls have less backpressure against them as you use them, and they feel "softer" than you will find at cruise speeds. The airplane will also be less responsive to the controls, having a "mushy" feeling as it responds slowly to control inputs. Your plane may have a tendency to wallow as you enter turns, the plane seeming not to want to respond at first to inputs, then moving slowly in the direction you are attempting to fly it toward. It will also take more movement of the control surface to cause the plane to reach a desired pitch or bank angle, as well.

While not as severe as speeds just above a stall, during takeoffs and landings the airspeed of the plane will be less than cruise speeds. The airspeeds will probably be slower during landings than takeoffs, depending on the manufacturer's recommendations. In these situations, you will not have the control authority available during cruise flight; but the controls will also not be as soft as during speeds very close to stalls. Here again, it will take more control movement to achieve a given bank or pitch angle as compared to cruise.

You should practice flying at these different airspeeds, building a sense for how the plane feels in each flight regime. After very little time, you will be able to approximate the airspeed you are flying, based only on how the controls feel. This helps to make you more of an instinctive pilot; and these instincts, combined with sound judgment and adhering to good piloting techniques, will make you a safer pilot. As always, if you are not proficient in these types of maneuvers, or feel uncomfortable executing them, take a qualified flight instructor along with you.

Slow Flight Entry

There are probably as many ways to enter slow flight as there are pilots. This section covers how I teach students to enter slow flight in a controlled, coordinated manner. Your plane may require different techniques; if that is the case, be sure to use the manufacturer's recommended procedures.

To begin, slow flight should always be done at altitudes that allow for safe recovery from inadvertent stalls or spins. This height varies based on the airplane and proficiency of the pilot; but slow flight should never be done below 1,500 feet above ground level (AGL). Pick an area that is clear of other aircraft and level off at normal cruise speeds as you prepare to enter slow flight. Once you are level and at a stable airspeed, you should make clearing turns to the left and right, verifying that no other aircraft are in your vicinity. These clearing turns can be from 90 to 360° turns in both directions, depending on the visibility from the plane and how busy the practice area is. I normally have my students perform 360° turns, just to get them in the habit of doing them.

Once these turns have been completed, pick a heading to fly as you enter slow flight. I normally have my students use a cardinal heading of north, west, east or south, using roads and fence lines on the ground that parallel the heading as external reference points. These external points to the left and right of the plane become more important as the nose of the plane blocks the forward view and prevents them from seeing ahead. Slow flight can be done with or without flaps, depending on the requirements of the lesson. To enter slow flight, I have the student reduce power to between approximately 1,500 to 1,700 rotations per minute (rpm), depending on the plane. When flying a plane equipped with a constant-speed prop, I use 15 to 17 inches of manifold pressure.

Heading and altitude should be maintained as power is reduced, the control yoke being pulled slowly back to reduce the plane's airspeed while maintaining altitude. If you intend to use flaps, once the airspeed is in the flap extension range, begin applying flaps, one notch at a time, until the flaps are extended the correct amount. Each time flaps are extended, there will probably be a tendency for the nose to pitch down, so be prepared to compensate for that. The airspeed will also be affected as flaps are added, their drag acting like a brake and slowing the plane more quickly. Also be prepared to add right rudder to overcome the left-turning tendencies that will become more prevalent as the plane slows. Not adding this rudder causes the plane to drift off the heading you want to keep.

Use the elevator to control your airspeed, increasing backpressure to slow it, and releasing backpressure to increase airspeed. Throttle settings will be used to control your altitude. If you are losing altitude, increase your power setting. Reduce the engine's power if the plane is gaining altitude. Some students find this correlation of elevator to airspeed and throttle to altitude difficult to grasp at first; but after very little practice, they become comfortable with the concepts. For instance, if you are losing altitude, increase the throttle setting. The increased power will cause the airspeed to climb, which necessitates that you increase elevator backpressure to maintain the correct airspeed. The now-greater lift produced by the wings will slow and stop the plane's descent once you have set the correct power settings.

As you approach the airspeed you want to hold, normally about 5 knots above stall speed, you will need to begin to add power again. This additional power lets you maintain a constant altitude and airspeed. How much power to add will depend on the plane you are flying and how it is loaded. It will take some practice to find out the correct power setting to be used; and once you learn the correct power for one plane, you will find other planes require a slightly different setting. The key to success in this case is to adapt to what the plane needs to hold altitude and airspeed.

As the plane becomes progressively slower, you will begin to feel the difference in how the controls feel. They become increasingly less responsive and require more input to make the plane move in the desired manner. You will notice that to hold the correct pitch angle, the elevator needs to have an increasing amount of backpressure until forces stabilize. The same is true of rudder because an increasing amount of right rudder needs to be added as the plane slows. As you hold the wings level, you will also notice it takes more aileron input to correct for forces that cause the plane to bank.

Once you are stabilized on heading and altitude, approximately 5 knots above stall speed, try some gentle banks. The plane will not roll into the turns crisply, but instead will be slow about its movements. It will begin to lose altitude almost immediately upon starting the bank. Recall that planes turn by taking some of the vertical lift component and use that lift to cause the airplane to turn. Figure 1-16 depicts the forces acting on a plane during a turn. As a result of the horizontal lift component using some of the total lift component, the plane has a reduced total lift. This causes the plane to begin to lose altitude as you enter turns, and it is why you must add backpressure to hold your altitude during a turn.

At very slow airspeeds, you may not be able to maintain your altitude using only additional backpressure on the control yoke. Due to the fact that you are just above stall speed, coupled with the fact that the stalling speed goes up in a turn, addition of this backpressure may cause the plane to enter a stall almost immediately. Adding power as you enter the turn will overcome the tendency to lose altitude and prevent you from stalling. You will again need to experiment with the plane you are flying to determine the correct amount of power that needs to be added. As the bank becomes steeper, you will find that even more power is required.

Use of rudder during turns to keep the ball centered also becomes critical. The high angle of attack, combined with power settings, can make the ball move a great deal to the right. Right rudder application will bring the ball back to the center. Note that during turns to the right, you may need to use a large amount of right rudder to keep the turn coordinated. To avoid potential spins if you inadvertently stall during turns, it is important to maintain coordinated flight throughout the turn.

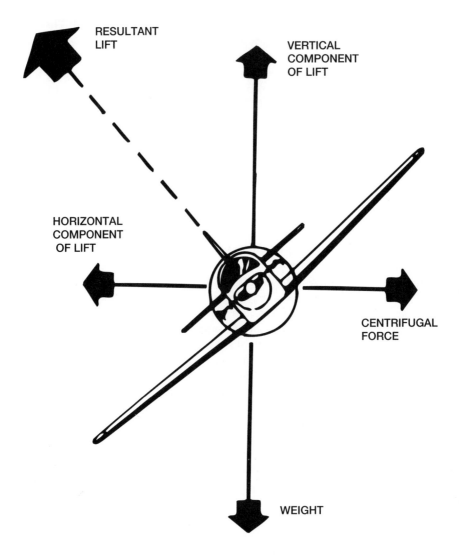

RESULTANT LIFT

VERTICAL COMPONENT OF LIFT

HORIZONTAL COMPONENT OF LIFT

CENTRIFUGAL FORCE

WEIGHT

Fig. 1-16 Lift components in turn.

Recovery from slow flight is accomplished in several steps. As you recover, you should continue to maintain the airplane's altitude and heading, avoiding altitude loss and unplanned turns in the process. First, power should be increased to cruise and the nose lowered as the power is brought in. The amount you lower the nose depends on the plane, but should be just enough to maintain altitude. The slower you add power, the slower you will lower the nose. The more quickly you add throttle, the faster you will need to push forward on the control yoke.

If your flaps are extended, the next step will be to retract them in stages. Depending on how many notches of flaps are extended, retract each notch, and then let the airplane gain airspeed and lift before retracting the next notch. As the flaps retract, the plane will have a tendency to settle and lose altitude due to the loss of lift they were generating. Be prepared by adding enough elevator backpressure to compensate for the loss of lift by increasing the angle of attack. After each notch of flaps comes up, you will then need to lower the nose again as the airspeed, and lift, increases. You can see that at airspeeds just above a stall, retracting flaps too quickly can cause the plane to stall or settle rapidly. Let the plane gain enough airspeed to avoid this prior to raising the flaps. If you handle these steps correctly, the plane will be at the same altitude and in cruise configuration as you complete the transition.

The more you practice entry into slow flight, keeping your altitude constant and your turns coordinated, the more comfortable you will become. Experiment with different flap settings and bank angles. At safe altitudes, I have my students enter steep banks that cause the plane to stall, even though they have not reduced their airspeed. This experience helps them gain a feel for the plane and understand how it reacts in this flight regime. After a short time, pilots can become very comfortable with slow flight as well as more competent in their ability to control the plane.

SUMMARY

This chapter covers a number of concepts related to stalls and spins. You need to know what a stall is, and what causes one. I know many pilots who are not at all comfortable with stalls, because of lack of knowledge and proficiency in the maneuver. This is where the real problem begins. One day, the plane you are flying will be at the fine line that divides slow flight from a stall. Knowing what causes a stall, how to avoid it, and how to recover from it will make all the difference in these borderline situations.

The early sections in this chapter covered a number of topics related to the definition of a stall, including how lift is generated, what the angle of attack is, the critical angle of attack, lift enhancement devices, and stall warning characteristics. Each of these topics is part of the foundation of knowledge you will need to fully understand stalls and spins.

The latter half of the chapter reviewed slow flight in detail. Among the areas of interest are control effectiveness during slow flight, left-turning tendencies as a result of P-factor and other physical characteristics, visibility during slow flight, the "feel" of slow flight, and entry into and recovery from flight at minimum controllable airspeeds.

Once again, before you attempt to practice any of the maneuvers described in this book, be certain you are proficient in them before executing them alone. Take a qualified flight instructor along who is comfortable teaching you the correct methods. Also, be certain that the plane is certified for each of the maneuvers. Chapter 3 covers spin certification for aircraft; but realize that if you ever attempt maneuvers that a plane is not rated for, you have just become a test pilot, in addition to breaking FAA regulations. Always use an aircraft that is certified for any maneuver you fly; the risk of doing otherwise is definitely not worth the potential outcome.

2

Stalls

CHAPTER 1 INCLUDES AN IN-DEPTH REVIEW OF SLOW FLIGHT AND reviews a definition of stalls. This chapter takes a more detailed look at stalls, covering a number of different stall scenarios in which a typical pilot should be interested. Among the topics included are normal stalls, accelerated stalls, approach stalls, departure stalls, and altitude loss during stalls. Discussion of each stall type includes entry into the stall from normal flight, slipping turns, and skidding turns. Within each of these topics is information about stall entry, stall recovery, and common errors pilots make as they execute these maneuvers. Altitude loss during stall recovery is also covered, in addition to factors that affect the stalling speed of the plane you fly. When you have completed the chapter, you should understand the various types of stalls a plane can enter and how to recover from them.

Ideally, you should not only read this chapter, but you should practice executing stalls as well. If you are not proficient with any of the maneuvers described, or are at all uncomfortable executing them, please find a qualified flight instructor to work with you as you practice each stall.

Note that under the proper conditions, executing stalls is as safe as performing any maneuver for which a plane is rated. I have talked to many students and pilots who do not like stalls for a variety of reasons. The high angle of attack, soft feeling of the controls, and deliberately dropping of the nose during recovery can be a new series of sensations for many pilots. The fact that many flight instructors do not like to teach full stalls seems to re-enforce the mindset that stalls are emergency maneuvers, straddling the line between flying and being out of control.

You should not be afraid of stalls. With the right training, in the right aircraft, stalls are like any other maneuver you fly. With practice, you can take a plane to safe altitudes, put it in a stall, and hold it in the stall for lengthy periods of time. I have done this with students, holding the plane's nose exactly on heading through the use of rudder pedals while the plane loses a thousand feet of altitude. Then I demonstrate how stall entry and recovery can be done, losing only a minimum of altitude and small pitch changes. With proper training, you can become as comfortable practicing stalls as you are with any other phase of flight training.

NORMAL VERSUS ACCELERATED STALLS

As you learned during flight training, stalls can be entered at any airspeed, from just above the stalling speed with flaps down, or Vso, through the remainder of the plane's flight envelope. For many pilots, stalls are viewed as a maneuver that takes place at slow airspeeds, the nose well above the horizon and the plane clawing at the air in an attempt to stay airborne. As noted in Chapter 1, stalls can take place at any airspeed, including speeds well above stall speed. Undoubtedly you have flown on bumpy, turbulent days when gusts of air cause the stall warning horn to go off as you cruise along at altitude. Figure 2-1 illustrates how even though the plane is moving along in level, forward flight, a sudden gust of air from below effectively creates a relative wind that generates an angle of attack that is beyond the critical angle. This increase in the angle of attack sets the plane up for a high-speed, or accelerated, stall situation. Unsuspecting pilots may never be aware of the true cause of their problems if they do not understand the aerodynamics behind a high-speed or accelerated stall.

Normal Stalls

This section takes an in-depth look at normal stalls. For the purpose of this book, normal stalls are those that do not require rapid control inputs or excessive g-loading to cause the airplane to stall. An example of an intentional nor-

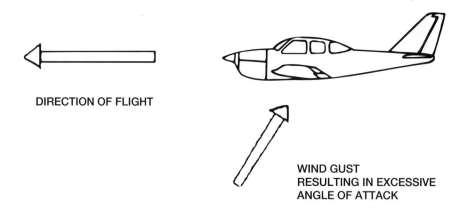

DIRECTION OF FLIGHT

WIND GUST
RESULTING IN EXCESSIVE
ANGLE OF ATTACK

Fig. 2-1 Wind gust induced stall.

mal stall entry would be to pull the engine power to idle, then ease the control yoke back slowly until the plane stalls. This is the type of stall taught to most private pilots during their flight training.

By contrast, accelerated stalls are more active in nature, the plane usually stalling at higher airspeeds than those associated with a normal stall. Accelerated stalls can take place as the result of rapid or excessive control inputs, highly banked turns, or severe turbulence. These stalls are more likely to take place when the pilot is making sudden, large control inputs, perhaps as a result of the need to suddenly maneuver in one direction or another. Accelerated stalls are covered after the review of normal stalls; but you should be aware of the difference between normal and accelerated stalls as you read each section.

Normal Stall Entry

Several variations of normal stalls exist, which are covered through the course of this chapter. Intentional entry techniques into normal stalls are similar for each of these variations. Entry into each stall is smooth and controlled, with differences dependent on the type of stall you want to execute. Before beginning stall practice, be certain to execute clearing turns in the practice area. Also, make sure you are at a safe altitude to allow for recovery with plenty of room to spare. For normal stalls, entry should be made from normal flight attitudes, the nose not pitching up too steeply as you enter the stall. Depending on the stall, you may execute it power off, or with power.

Power Off Stall Entry

In preparation for power off stall entry, once you have established the airspace is safe for stall practice, slowly ease the throttle to idle. There are several reasons for this slow reduction of power. First, I have witnessed too many students spending fifteen or twenty minutes flying out to the practice area, climbing during most of the trip. As a result of the nose high climb attitude, the temperature of the engine can rise. Arriving at the practice area, the pilot executes a couple of clearing turns, yanks the throttle back to idle, stalls the airplane, shoves the nose down for recovery, and finally slams the throttle forward to the firewall again.

This type of stall practice is very hard on air-cooled aircraft engines. During the climb the engine heats up, and then it is rapidly cooled as the throttle is quickly closed and the nose dropped, increasing the cool air flowing over the engine. Engine treatment like this can result in the cylinders "shock cooling," the rapid drop in temperature causing the cylinders to crack. Repeated cycles of heating and rapid cooling can reduce the life of the engine. For this reason, it is best to slowly cool the engine after the climb to altitude by slowly reducing the throttle to idle.

As power reaches idle, the control yoke should be smoothly pulled aft to its full travel, the nose of the plane rising in the process. At this point, airspeed should be bleeding off and the plane approaching stall speed.

Power On Stall Entry

As the name of this section implies, it covers how to enter intentional power on, normal stalls. Power on stalls are often referred to as departure stalls, with the engine developing full power and the airspeed falling off as the plane slows while in a climb. Normal entry into power on stalls involves clearing the practice area, just as in the power off stall, prior to executing the stall. Once this has been done, from level flight the plane is slowed to the correct airspeed. This reduction in speed can be accomplished by reducing the engine's power, and then easing the control yoke back until the plane slows to the desired speed. At this point, full power is added as the throttle is smoothly pushed forward to the firewall, and the nose raised to maintain the desired airspeed.

Another method for reducing airspeed prior to adding full power is to raise the nose of the plane from level flight, with the throttle at cruise settings. As the nose comes up, the airspeed will bleed off. Once the target airspeed is achieved, add full power; the airplane is now in essentially the same configuration as in the prior example. In most cases, the airspeed used for departure stalls will be either V_x, the best angle of attack, or V_y, the best rate of climb, depending on the aircraft manufacturer's airspeed recommendations for the maneuver you are going to execute.

I normally teach the first stall entry technique to students. They seem to have an easier time getting airspeeds and power settings correct as compared to the second technique. You may have learned other variations, so use the techniques that work best for you. The main point to keep in mind is that you fly the airplane through the entire maneuver and control the airspeeds and attitudes as you execute the maneuver.

ACCELERATED STALLS

If you have ever pulled back hard on the yoke in a steep turn and felt the airplane rumble around you, you have felt the onset of an accelerated stall. If you had the misfortune of pulling the control yoke back even further, and the nose snaps down or the wings suddenly depart in another direction, you have just experienced an accelerated stall. Under the right conditions, accelerated stalls are not inherently more serious than a normal stall, but those are a limited set of circumstances.

Accelerated stalls cause greater stresses on a plane than normal stalls, the result of higher g-loadings that are imposed on it. This is one of the reasons that aircraft manufacturers placard airplanes with maneuvering speeds. At or below the maneuvering speed for a plane, abrupt control inputs are much less likely to damage the airplane's structure. The plane will normally stall before damage is done. For the same reason, when flying into rough or turbulent air, you should reduce your airspeed to maneuvering speed. When at maneuvering speed, the plane will normally stall before turbulence can cause damage to the plane. Flying near, but at or below, maneuvering speed provides a safer airspeed margin to help you avoid potential stalls. Above the design maneuvering speed, excessive control inputs can overload the structure of the plane, causing permanent damage or failure of the aircraft. For this reason, accelerated stalls should NEVER be attempted above maneuvering speed.

Figure 2-2 depicts the load factor, or g's, a plane encounters as the angle of bank increases while the plane maintains a constant altitude. The graph documents that as the angle of bank increases, the g-load begins to increase exponentially, reaching 9 g's in an approximately 85° bank. There is a sharp rise from 5.76 g's in an 80° bank to 9 g's in just 5 more degrees of bank.

To figure out why this is a problem, look at the load limit factors that aircraft are designed to. Figure 2-3 shows the three categories under which general aviation planes are most frequently certified. First is the normal category.

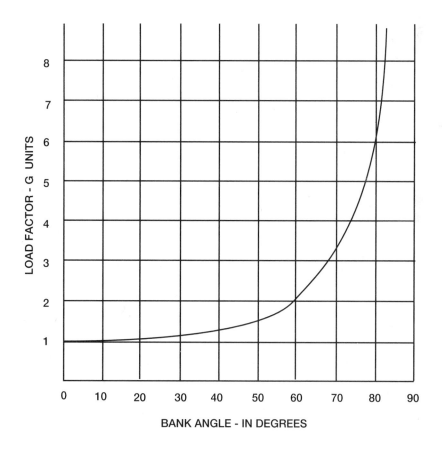

Fig. 2-2 Bank vs. load factor.

CATEGORY	LIMIT LOAD FACTOR
NORMAL	3.8
UTILITY	4.4
AEROBATIC	6.0

*TO LOAD LIMITS GIVEN, A SAFETY FACTOR OF 50% ADDED

Fig. 2-3 Aircraft category load factors.

Looking at the right column, you can see that the maximum positive load limit is 3.8 g's for normal category planes. Suddenly the 85° bank and its 9 g's seem to raise a warning flag. Even with the 50 percent safety factor that is designed into an airplane, this bank will produce g loadings well above the load factors that the plane can safely sustain.

Figure 2-4 shows why load factors increase in a turn. "In a constant altitude coordinated turn the load factor (resultant load) is the result of two forces: (1) pull of gravity, and (2) centrifugal force" (*Pilot's Handbook of Aeronautical Knowledge*, p. 24). In the figure you can see that in a steep banked turn, the centrifugal force component is greater than in a medium banked turn and the resultant load limit is also higher. Now the reason the g-loads increase as the bank angle steepens becomes easier to understand.

Aircraft certified in the utility category have a maximum load limit of 4.4 positive g's, still short of the loads imposed in only an 80° bank. Those aircraft certified in the aerobatic category have a maximum load limit of 6.0 g's. Even highly stressed aerobatic aircraft are restricted in load limits that may be imposed on them. As an aerobatic flight instructor, I have had students regularly pull too hard on the control stick at speeds below those necessary to enter the maneuver they are attempting, but still well above stall speed. The plane shudders and the wing breaks to one side or the other as the bewildered student attempts to level the plane. This is an accelerated stall.

When you are flying at cruise speeds, the wings are capable of producing more lift than at slower airspeeds. The higher the speed at which you are flying, the more lift capability the wings have. However, this additional lift capacity also means that the plane will be able to support additional weight. This additional lift capability, and its accompanying higher stall speeds, allow you to place higher load factors on the plane. Not only does excessive turbulence have the potential to cause structural damage, but so does overuse of the controls in recovery from stalls and spins.

Recovering from a stall or spin can normally be accomplished with relatively small inputs to the elevators. I have had students who are uncomfortable with stalls shove the control yoke forward to the stop during recovery, attempting to get the nose down as quickly as possible after the stall takes place. There

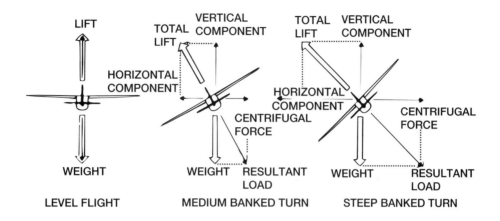

Fig. 2-4 Factors acting on aircraft during a turn.

have been those who, once the nose is down and airspeed is building, have yanked back hard on the elevators as they try to get back to level flight as quickly as possible. I tend to see this more frequently from students in spin recovery, due to the greater nose down attitudes found in that maneuver. These situations are prime candidates for accelerated stalls. The plane's airspeed is building and a sudden nose up input on the elevator can produce the high-speed buffet associated with the onset of an accelerated stall. Additional back-pressure can cause the plane to stall and impose potentially damaging loads on the plane's structure.

To avoid excessive g-loads, the *Pilot's Handbook of Aeronautical Knowledge* (p. 27) recommends:

1. Operate the airplane in conformance with the *Pilot's Operating Handbook*.
2. Avoid abrupt control usage at high speeds.
3. Reduce speed if turbulence of any great intensity is encountered in flight, or abrupt maneuvers are to be performed.
4. Reduce the weight of airplane before flight if intensive turbulence or abrupt maneuvering is anticipated.
5. Avoid turns using an angle of bank in excess of 60°.

Common Errors

There are a number of common errors that tend to precipitate accelerated stalls. If you are aware of these factors, and work to avoid them, you will help increase your chances of avoiding unintentional accelerated stalls. After reading the material just presented, you probably already can think of several ways that pilots get themselves into unexpected accelerated stalls. One of the most common is to fly through turbulent air above the design maneuvering speed and a sudden upgust of air causes the plane to stall. Knowledge of the flight conditions, including any turbulence forecasts, can help avoid this situation. Knowing the maneuvering speed of the plane you fly, and flying at or below that speed when in turbulent conditions, can also help avoid problems.

Steeply banked turns are also common methods for entering accidental accelerated stalls, with the plane stalling as the bank becomes too steep and the flight loads become greater than the available lift. Excessive elevator backpressure at speeds above the maneuvering speed can also cause accelerated stalls.

To summarize, this section reviews the differences between normal and accelerated stalls. Normal stall entry is very controlled and produces little, if any, additional g-loads on the plane other than those found during normal flight. Accelerated stalls are usually entered at higher speeds than the normal stall and have the potential of producing damaging loads on the plane if excessive airspeed or control movements are involved.

APPROACH STALLS

Approach stalls are entered from airspeeds, flap and gear configurations that match those of the plane during an approach to landing. In this section we are going to cover several different versions of approach stalls, including normal approach stall entry, entering into approach stalls from slipping turns, and entry into an approach stall from a skidding turn. By understanding how to enter and recover from approach stalls, you will be better prepared to avoid them in actual approach situations. Since most aircraft stall accidents happen during approach and departure, it is critical that you are aware of what causes an approach stall.

Normal Approach Stall Entry

Normal entry to an approach stall is a very controlled procedure. Prior to beginning any stall practice, be sure to review the procedures recommended by the operations manual for the airplane you are flying. Climb to a safe altitude prior to beginning stall practice. Once you reach altitude in the practice area, execute clearing turns to be certain no other aircraft are present, being sure to check below you. Pilots frequently forget this particular step until they get in the habit after several stalls. In an approach stall, you want the airplane in the landing configuration. For example, if you are flying a retractable gear airplane, the gear needs to be extended during the stall. The same is true for flaps, if your plane is equipped with them. This means that you will probably need to slow the plane down before lowering the gear and flaps. To do this, reduce power to somewhere in the 1,500 to 1,700 rpm range, or manifold pressure to the 15- to 17-inch range. Holding the plane at a constant altitude, allow the airplane to slow, extending the flaps and gear at the appropriate airspeeds.

Once the plane is in the correct approach configuration, you are ready to begin stall entry. Adjust the power to approach settings, which is idle for many general aviation planes. If your plane requires carburetor heat when the engine is at low-power settings, apply carb heat prior to reducing engine power. Please note that carb heat should be added before reducing engine power. Most carburetor heat systems are designed to take outside air and pass it through a muffler shroud, heating the air on its way to the carburetor. The muffler cools after the power is reduced. Waiting to add carb heat until after power has been reduced may give the muffler sufficient time to cool so that the air will not be heated enough to melt the ice in the carburetor. Once power is set, the airspeed is reduced to the approach speeds normally used during the descent. At this point, you should be in a normal approach descent. I have students practice approach stall entry both from straightahead glides and while making descending turns. This gives them the opportunity to experience the plane's reaction in both situations and allows them to compare the differences between the two.

Straightahead Glide

During the descent, begin to ease the control yoke back a selected heading. The airspeed will begin to bleed off as the nose of the plane begins to rise. As it does, the noise from the engine and slip stream will change, in addition to the controls becoming softer. Eventually the speed will deteriorate to stall speed, the control yoke most likely in the full aft position. Keep the ball centered using rudder pedals as you pull the yoke back. It will take progressively more rudder to do this as the plane's nose rises and the airspeed decays. When the plane stalls, the nose will probably drop perceptibly. How much the nose drops, or breaks, varies from plane to plane. At this point in the maneuver, you should smoothly release elevator backpressure on the control yoke and lower the plane's nose. You will then need to add full engine power.

To avoid excessive altitude loss during the maneuver, do not force the nose of the plane too far down during recovery. Normally, just easing forward slightly on the elevators is enough to break the stall. In addition to avoiding unwanted altitude loss, this use of relatively small inputs also makes recovery from the stall smoother. You will not find yourself suddenly up against the seatbelt in a weightless environment, with dirt from the floor, lost pens, and other articles inside the plane floating around your head as the plane dives for the ground.

As the plane accelerates to safe flying speeds, ease in nose up elevator to bring the plane back to level flight. If the plane tends to drop a wing during the stall, use rudder, not ailerons, to pick up the low wing. Use of ailerons to level the wings at too slow an airspeed, or while still in the stall, can aggravate the stall situation. Most wings on today's general aviation aircraft are designed with washout. Figure 2-5 depicts the differences in the angles of attack between the wing root and the wing tips. This "twist" in the wing causes the wing root to stall before the outer portions of the wing. As a result of this design feature, the ailerons remain effective longer after the wing begins entry into a stall, which provides additional control of the aircraft during a stall.

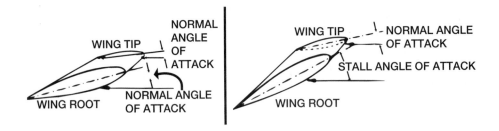

Fig. 2-5 Wing root vs. wing tip angle of attack.

However, it is quite possible to further complicate the stall if excessive aileron input is used to hold the wings level. "For example, if the right wing dropped during the stall and excessive aileron control were applied to the left to raise the wing, the aileron deflected downward (right wing) would produce an even greater angle of attack (and drag), and possibly a more complete stall at the tip as the critical angle of attack is exceeded" (*Flight Training Handbook*, p. 146). This is less likely to take place if rudder is used to keep the wings level during a stall. However, if you do use aileron, make sure you keep the inputs coordinated and also use rudder to keep the ball centered. If you do not, the adverse yaw created by the lower aileron may cause the plane to enter into a spin. "A spin will not occur if directional (yaw) control is maintained by timely application of coordinated rudder pressure" (*Flight Training Handbook*, p. 146).

For this reason the FAA states that "the primary use of rudder in stall recoveries is to counteract any tendency of the airplane to yaw, or slip. The correct recovery technique then, would be to decrease the pitch attitude by applying forward elevator pressure to break the stall, advancing the throttle to increase airspeed, and simultaneously maintaining direction with coordinated use of aileron and rudder" (*Flight Training Handbook*, p. 146).

Figures 2-6 and 2-7 are a series of photographs showing the attitudes that an airplane passes through during an approach to landing stall. The sequence in Figure 2-6 shows the view you would see from inside the cockpit, from the pilot's seat on the left side. The horizon drops as the nose is raised, making it more difficult to see ahead. Once the stall takes place, the nose is lowered enough to break the stall, but not an excessive amount. In the forward views, Figure 2-6a demonstrates how little the nose is raised above the horizon as the plane approaches stall speeds. And, more surprisingly to many pilots, Figure 2-6b is the amount of nose down pitch that is necessary to recover from the stall. You do not need nearly the forceful push on the control yoke that some pilots have been taught to use during recovery.

The sequence in Figure 2-7 shows the same stall, but looking out the left window of the plane. This perspective more clearly shows the angles of attack that the plane goes through during entry into the stall and recovery from it. This particular view often surprises pilots. They sometimes associate stalls with

Fig. 2-6a Approach to landing stall entry: forward view.

Fig. 2-6b Approach to landing stall recovery: forward view.

Fig. 2-7a Approach to landing stall entry: left side view.

Fig. 2-7b Approach to landing stall recovery: left side view.

very high angles of attack, and this is not always the case. This series also indicates how little the angle of attack needs to be reduced to break the stall. Finally, the plane is once again in level flight. Both of these sequences were executed with the flaps retracted and power reduced to idle.

You should practice straightahead gliding approach stalls with different aircraft configurations, as long as they fall within the limits prescribed by the aircraft's manufacturer. Try the stall with different flap setting, from no flaps at all to full flaps. If you are flying a retractable gear airplane, practice the stalls with the gear extended and retracted. While you are doing this, notice the differences in how the plane stalls. What was the pitch of the plane when it stalled, the airspeed, how did the controls feel? Be aware of the plane as you fly.

Turning Approach Stalls

This section covers three areas related to turning approach stalls. The first example is a normal, coordinated turn during which you slow the airplane until it stalls. The second covers severe slips during a turn and how they affect the stall. Finally, severe skidding turns leading to a stall are reviewed. As you will see, coordination between the ailerons and rudder become very important during turning approach stalls. Each discussion is accompanied by a series of photographs to give you an idea of what the stall sequence looks like from the pilot's seat. Note that I do not recommend practicing the skidding and slipping turn stalls without a very strong understanding of spin recovery. The nature of a stall during both of these scenarios sets you and the plane up to enter a spin if you do not recover correctly. For this reason, you should only practice these maneuvers in a spin-rated airplane.

Normal Gliding Approach Stalls

Normal turning approach stalls are made like any other power off stall, with the power at idle, the controls coordinated, and airspeed bled off by easing back on the control yoke. The difference from the straightahead gliding stall is that the plane is in a turn during the maneuver. This is going to affect how the plane reacts during the stall, and how the controls will be used as the plane slows and eventually stalls. Due to the changing airspeeds and the desire to hold a constant bank, you will constantly need to adjust the amount of control input, including the ailerons and rudder in order to maintain the plane in coordinated flight. During this maneuver, some pilots have a tendency to enter either a slip or a skid, leaving themselves open for a spin after the plane stalls. The plane will also have a stronger tendency to break, dropping a wing as the plane stalls. As discussed earlier, excessive aileron use during stall recovery can aggravate the stall. Correct stall recovery control techniques become very important in these stall situations.

In this sequence, you begin with a climb to a safe altitude and clearing turns in the practice area to ensure that the area is clear of other aircraft. Set the aircraft up in a landing descent, with the flaps and gear in the standard configuration.

With the power at idle, begin a turn to the right or left, keeping the ball centered. Use a medium bank, no greater than 30°, just as you would in the pattern. Then, while maintaining the bank angle, ease the control yoke back and begin to bleed off airspeed. As the plane slows, you need to be aware of the plane's reaction—for example, if it begins to drop a wing, slip, or skid. As the plane continues to slow, it will require larger control inputs to hold the correct attitude. Finally, it will begin to buffet, then stall. When the stall takes place, reduce backpressure and bring in full throttle, being sure to use the rudder to keep the ball centered during the break. Then use small aileron inputs to level the wings.

Figures 2-8 and 2-9 are pictures showing how this stall sequence looks. Figure 2-8 shows the forward view and Figure 2-9 shows the view out the left window of the plane. Once again, you can see that pitch angles during stall entry and recovery do not need to be excessive in order for the stall sequence to be successful.

Common Errors Pilots can make a number of common errors when practicing approach stalls. The first is failing to keep the ball centered as the plane slows during the turn. As the angle of attack and control authority changes, so will the amount of rudder needed to maintain coordinated flight. You should also avoid excessive use of ailerons to control the bank angle of the wings. This

Fig. 2-8a Turning approach stall entry: forward view.

Fig. 2-8b Turning approach stall recovery: forward view.

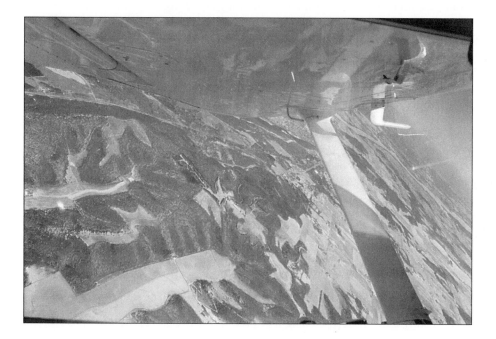

Fig. 2-9a Turning approach stall entry: left side view.

Fig. 2-9b Turning approach stall recovery: left side view.

can cause an even more complete stall along the wing. Another common error is that pilots do not maintain enough backpressure on the control yoke to slow the airplane into a stall. Instead, they only slow the plane, which increases the rate of descent, but do not get the plane to stall.

Another error pilots make is to either not recover from the stall by reducing elevator backpressure, or overcontrol the amount the nose drops after the stall by pushing too far forward on the control yoke. Both situations should be avoided. In the first, the plane may enter a secondary stall, increasing the amount of altitude the plane loses before recovery is completed. In the second, the plane once again loses excessive altitude during the recovery due to the dive induced by the nose down attitude. In an ideal stall recovery, the nose of the plane drops only enough to break the stall.

Pilots sometimes also fail to bring in full throttle as they recover from the stall. Not adding throttle increases the amount of altitude lost during the recovery. In a low altitude stall situation, not adding full engine power during the recovery may turn a salvageable situation into a serious problem. In a stall recovery, you should advance the throttle smoothly to full power. Avoid shoving the throttle in too quickly, as this may cause some engines to falter as a large volume of avgas is introduced into the cooled engine.

In a textbook stall recovery, the nose of the plane drops just enough to restore sufficient lift to the wings while the engine is smoothly advanced to full power. When done correctly, I have seen no perceptible altitude loss from a stall in some aircraft. While this is not possible in all planes, the minimum amount of altitude loss is one criterion you should watch as you practice.

Slipping Approach Stalls

Before reviewing slipping approach stalls, you should understand the differences between slips and skids. In each case, the plane is in an uncoordinated situation, with the rudder and ailerons not working in concert with each other. Slips are not necessarily a bad maneuver and, in fact, are very useful in some landing situations.

A skid takes place when a plane has insufficient bank for a given radius of turn (*Flight Training Handbook*, p. 42). Put another way, the pilot is probably trying to force the plane to yaw in the direction of the turn with excessive rudder, and the wings of the plane remain close to a level attitude. During the early days of aviation, most turns were skidding, because pilots of that time period believed that the plane might get out of control if put in a bank. In those days, pilots made wide, skidding turns, using rudder to force the nose of the plane around during the turn. Today, proper flight training teaches that banks should be used when making turns.

Pilots that get into skidding turns often do so during landing. The most frequent setup to these skidding turns is when a pilot misjudges the turn from base to final, beginning the turn too late. Knowing that the stall speed goes up in steep turns, and not wanting to get into a steep bank close to the ground, they begin to push very hard on the rudder pedal, yawing the plane in the direction of the turn. Ailerons may be held either in their original position or used to hold the wings of the plane level. In severe cases, the plane stalls and one wing suddenly drops, or the plane enters a spin. In an aerobatic routine I fly, I make a 180° turn holding the wings level and using only rudder to turn the plane. While this looks good for an aerobatic maneuver, it is not the proper way to fly a turn to final.

A slip is a descent with one wing lowered and the longitudinal axis at an angle to the flight path (*Flight Training Manual*, p. 102). A slip is useful in two situations: to lose excess altitude without a buildup in airspeed and to counteract a crosswind. Prior to flaps being incorporated on most aircraft, pilots learned to use what is known as the forward slip (Figure 2-10) to allow the plane to make a steeper descent without diving at the runway. The forward slip is entered by lowering a wing and at the same time inputting opposite rudder to the point that the airplane's longitudinal axis is at an angle to the original flight path. The amount of yaw is dependent on how much bank was initiated and should be enough to maintain the original ground track (*Flight Training Manual*, p. 102). This has the effect of yawing the plane, presenting the side of the fuselage to the relative wind and causing a great deal of drag. This drag increases the rate of descent without a substantial increase in airspeed.

When in a side slip, the longitudinal axis remains on the original heading and the flight direction changes based on the amount of bank you enter. By side-slipping into the wind, you can keep the plane flying down the center of

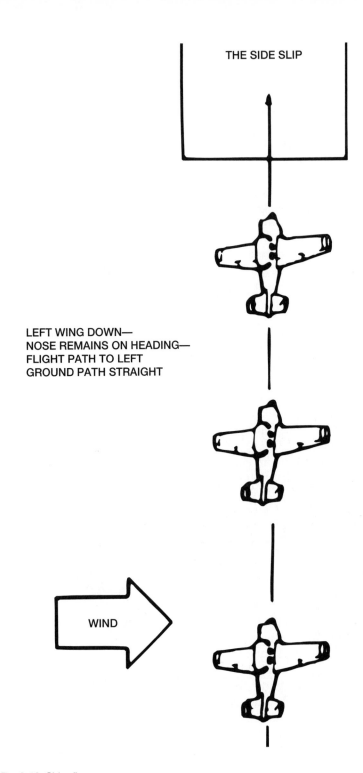

THE SIDE SLIP

LEFT WING DOWN—
NOSE REMAINS ON HEADING—
FLIGHT PATH TO LEFT
GROUND PATH STRAIGHT

WIND

Fig. 2-10 Side slip.

the runway, not allowing it to drift with the wind. The landing gear remains parallel to the direction of flight, and will not touch down on the runway at an angle. This avoids the potential of skidding and other handling difficulties that can take place when a plane lands pointing in a direction other than the one it is moving. The application of slips during landing is left for other books, but you need to understand what slips and skids are.

Figures 2-11a and 2-11b show the differences between a slip and a skid as indicated by the turn and bank indicator. Figure 2-11a shows a turn to the right, the ball slipping down and indicating the pilot needs more right rudder to maintain coordinated flight. Figure 2-11b shows a plane making a turn to the right, but this time making a skidding turn. In this case, the pilot needs to use less right rudder in order to maintain coordinated flight. There is a rule of thumb: "Step on the ball." When making turns, if the ball is over to the left of the tube, you need to push left rudder; if it is to the right, push right rudder. You should keep in mind that adding or reducing your rudder inputs affects your bank angle and how much aileron you need to maintain the angle of bank you need to fly. This means that as you change rudder, you will also need to change the amount of aileron to maintain a constant bank.

Now that you have reviewed the differences between slips and skids, let's cover what a stall from a slipping approach turn is like. The discussion assumes

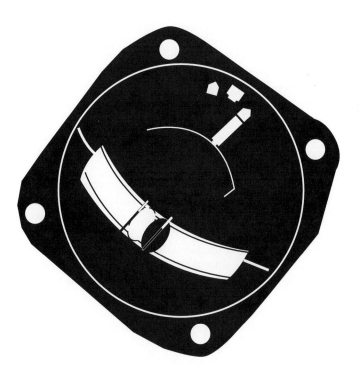

Fig. 2-11a Turn and bank indicator: slip.

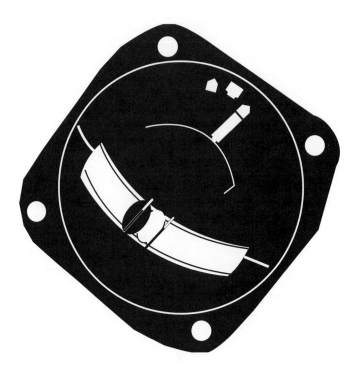

Fig. 2-11b Turn and bank indicator: slide.

you are following recommended procedures for entry into the practice area, including clearing turns. The entry into the approach stall is the same as covered in the section "Normal Approach Stalls." Once the plane is in the descent, begin putting it into a slip while easing back on the elevators. For example, if the plane is turning to the left, smoothly add right rudder, using ailerons to hold a constant bank angle. While it varies from plane to plane, you may find the plane you are flying has a strong tendency to roll in the direction of the rudder as you add it.

Continue to add backpressure and rudder. At some point, you will notice the plane does not "feel right." The slip will try and "push" you toward the low wing. The balanced feel the plane has during a coordinated turn is now lost. Stall buffeting is likely to begin at a higher airspeed than was previously seen as a result of the excessive drag the fuselage generates while in the slip.

During slips, especially at slow airspeeds, the airspeed indicator may have a significant error in it. In many airplanes, when slipping to the left, the airspeed indicator shows a lower airspeed than the plane is actually flying. To the right, it may register higher. This is due to the change in pressure at the static pressure point as the fuselage is exposed to the slipstream. The pressure differential between the pitot tube and static source are used to display the correct airspeed on the airspeed indicator. Because the airspeed indicator is calibrated for a specific pressure differential between the pitot tube and static source, when the plane slips this pressure difference changes and aggravates the erroneous airspeed indication.

As the plane stalls, it can easily drop a wing and enter a spin. Be sure to immediately input rudder to center the ball and reduce elevator backpressure. Smoothly add full throttle, and then level the wings once safe flying speed has been achieved. Whenever the plane stalls in uncoordinated flight, the chances for spin entry are increased. Spin recovery is covered in later chapters of the book, but you should avoid a spin when practicing this maneuver.

Figures 2-12 and 2-13 show the forward and side views from the plane during a slipping approach stall. These sequences were performed in a Cessna 152. While different aircraft models will react differently, the views show what can take place in a severe slipping approach stall. In this example, the plane was in a bank to the left, with full right rudder and the elevator in the full aft position. A constant bank of approximately 30° was held through the turn. The plane buffeted heavily during the stall, but did not enter a spin. Recovery was initiated by reducing the angle of attack and coordinating rudder with the ailerons. Full power was applied as the wings were rolled to level and the nose was eased slightly below the horizon. Once safe flying speed was achieved, the plane was brought back to straight and level flight.

While the plane would not enter a spin during the approach stall in a slip with power off, if power was added during the stall, the nose rose dramatically and the plane then entered a spin to the right. In this case, the sequence of stall recovery steps becomes very important. Those pilots who have been taught to push the throttle in immediately upon stall entry will be surprised as the plane reacts by entering a spin. My hypothesis is as follows: in a slipping approach, stalling the elevators and rudder of a plane does not have sufficient authority to force the plane into a spin that is opposite to the direction of the turn. Once throttle was applied in the example, the slip stream increased and the control surfaces were then able to overcome the forces of the turn. This caused the plane to enter a spin.

Once again, other planes may react very differently from the Cessna 152 in a slipping approach stall and may very easily enter a spin as the plane stalls. You should be prepared for a spin if you practice this maneuver. For that reason, be sure your plane is spin-certified and you are proficient in spin recovery, or have a flight instructor along who is. As you can see from the pictures, the plane's nose is very high for power off flight. The slip literally pushes you to one side of the aircraft and the buffeting is a strong warning that the plane is in an imminent stall. While it seems impossible for an alert pilot to get into this situation during an approach to landing, if you do not constantly fly the plane, or become distracted during the approach, you could find yourself in this slipping stall. Avoid the situation entirely by being aware of the plane and flying it at all times.

Common Errors All of the common errors covered in normal approach stalls apply to the slipping approach stall. Correct use of the rudder and

Fig. 2-12a Slipping approach stall: forward view.

Fig. 2-12b Slipping approach stall: forward view.

Fig. 2-13a Slipping approach stall: left side view.

Fig. 2-13b Slipping approach stall: left side view.

ailerons becomes critical in avoiding spin entry or excessive loss of altitude during recovery. Some pilots are surprised by the rapid break that takes place as the plane stalls in the uncoordinated flight maneuver. As this happens, there is a tendency to overcompensate, aggravating the stall condition. Use the smallest control inputs necessary to recover from the stall; this will help you avoid making the situation more difficult.

Skidding Approach Stalls

Entry into skidding approach stalls is much like the slipping approach stall. Begin a normal approach stall entry; but as the plane slows, use increasing amounts of rudder in the direction of the turn. For example, if you are turning to the right, begin to add an increasing amount of right rudder, but do not let the angle of bank increase. This may require that you use opposite aileron to hold the bank angle constant. As you add rudder, increase the amount of elevator backpressure, slowing the plane toward a stall.

The ball of the turn and bank indicator will skid to the outside of the turn, with centrifugal force throwing it, as well as you and your passengers, outward. As you continue to add more rudder pressure, the plane will take on the characteristic feel of an uncoordinated turn. With a little practice, you can tell just by the feeling in the "seat of your pants" that the plane is not coordinated. Many pilots I have given aerobatic flight instruction to are slaves to the instruments of the planes they fly. When first beginning to learn aerobatics, they must adjust to the plane's lack of instruments. You should practice making turns without looking at the ball continuously to see if your turn is coordinated. Soon, you will be able to tell how the turn is going by the feeling you have. You will feel yourself sway to one side or the other during the turn and begin to compensate without even thinking about what you are doing. This is the mark of a pilot that has a good feel for the plane.

As the plane stalls, it will once again break, with a strong probability of one wing dropping. The break may be more rapid than in the normal approach stall, depending on the design of the plane. Release backpressure and coordinate through the proper use of rudder. Then add throttle and level the wings, being sure not to use excessive ailerons to level the wings.

Figures 2-14 and 2-15 show the forward and side views from the plane during a skidding approach stall. Like the slipping turn, this maneuver was executed in a Cessna 152. In this stall sequence, the plane's wings were held in a wings-level attitude, but full left rudder was applied as the nose was raised with increasing backpressure on the control yoke. This had the effect of yawing the plane rapidly to the left and slowing the airplane. Right aileron input became necessary to keep the angle of bank from increasing, further complicating the cross-control situation. This is a classic entry into a spin and the plane reacted accordingly by yawing to the left, then entering a spin.

The photographs document how the nose drops slightly as greater amounts of left rudder are applied, while increasing opposite aileron is necessary to hold the wings level. Finally, the plane's left wing drops, and continues into a spin

Fig. 2-14a Skidding approach stall: forward view.

Fig. 2-14b Skidding approach stall: forward view.

Fig. 2-14c Skidding approach stall: forward view.

Fig. 2-15a Skidding approach stall: side view.

Fig. 2-15b Skidding approach stall: side view.

Fig. 2-15c Skidding approach stall: side view.

to the left. While all airplanes may not enter a spin from this flight configuration, the potential exists while the airplane is flown in this manner. A skidding turn 300 feet from the ground to final is not the place to find out how proficient you are at spin recovery.

Common Errors Those errors discussed for both normal approach stalls and slipping approach stalls apply to skidding approach stalls. Be prepared for the differences in how the plane reacts during the skidding turn as compared to the normal approach stall and the slipping approach stall.

DEPARTURE STALLS

As the name implies, this section discusses departure stalls. Departure stalls simulate the plane being in takeoff configuration and how it reacts during a stall. As the departure stall is entered, the engine is normally at full throttle with the plane in an increasingly steep climb. These stalls can be done with the plane climbing straightahead, or in a turn. Both straightahead and turning departure stalls, including both slipping and skidding turns during the stall, are reviewed.

Like all stalls, departure stalls should be done at a safe altitude in an area clear of other aircraft. If you are unfamiliar with the maneuvers, or lack proficiency in them, be sure to bring along a qualified flight instructor to increase your learning opportunity and safety during the maneuvers. The plane you are practicing in must be certified for these types of stall.

Whenever you fly, and in particular when practicing stalls and spins, make sure the plane you are flying is within the manufacturer's weight and balance guidelines. When a plane falls outside of the weight and balance limits, it may be difficult, if not impossible, to recover from a stall or spin. For instance, if you are flying a plane with a center of gravity (CG) past the rear of the envelope, once the plane enters a tail-low attitude for the stall entry, you may not have the elevator authority at slow airspeeds to recover. This condition can become even worse in a spin, a topic covered later in the book during discussions of flat spins.

Normal Departure Stall

As just stated, a departure stall is executed while the plane is in a normal takeoff configuration: the engine developing full power and the airplane in a normal climb using a standard climb speed. From that point, the plane's nose is progressively raised by increasing elevator backpressure on the control yoke. If the plane is in a bank, the angle should be held constant; and if it is in a wings-level attitude, the heading should remain constant. In either situation, the amount of rudder necessary to maintain coordinated flight will increase as the airspeed slows. In a turning departure stall, the amount of aileron input needed for a constant bank will also change.

Because the engine is developing full power, the angle of attack during the stall will be greater than those experienced during power off stalls. Depending on the amount of horsepower the plane you are flying produces, this could result in a considerable increase in the pitch angle before the plane stalls. As the plane approaches stall speed, buffeting should set in. Buffeting may be much stronger with the engine producing power than you felt during the power off stall maneuvers. The higher angle of attack, combined with the additional energy that the air flowing over the wings generates, can create stronger turbulence, and therefore, more pronounced buffeting.

As the plane stalls, the nose will break, possibly much more quickly than in a power off stall. Smoothly release backpressure on the elevators, allowing the nose to drop to just slightly below the horizon. Avoid the tendency that some pilots have of forcefully shoving the nose of the plane down, thinking that this will more quickly recover from the stall. In fact, this increases the amount of altitude lost during stall recovery, with the plane diving toward the ground. The engine should already be producing full power; but if it is not, be sure to ease the throttle in to full power. Here is a word of caution: if you have any up elevator in as you add power during any stall recovery, the nose of the plane will probably rise. To avoid pitching up suddenly and potentially entering a secondary stall, you may need to gently ease the control yoke forward as you add power, thus reducing nose up tendencies.

During the stall and afterwards, you should keep the ball centered with correct use of rudder. While it seems this point is being repeated very often in the book, many pilots do not understand the importance of coordinated controls. As we saw in the skidding approach stall section, uncoordinated use of rudder set the airplane up for spin entry. It is an unfortunate fact that some pilots today fly with their feet flat on the floor, not even touching the rudder pedals while they fly. Don't be one of them. Use the rudder in a coordinated manner and practice keeping the ball centered.

Once a safe airspeed has been achieved after recovery, the plane can be flown back to straight and level flight and the power reduced to normal levels. Figures 2-16 and 2-17 show the now-familiar forward and side views of a standard departure stall in a turn to the left. As you can see, the angle of attack is greater than those of the approach stall. The nose rises as the airspeed slows, which can be seen in both views. As the plane stalls, the nose drops and is positioned just below the horizon during stall recovery. While there is a greater change in pitch from stall entry to recovery, the amount the nose drops below the horizon is not excessive in this example.

As you practice departure stalls, you should vary the amount of flaps you are using, from flaps fully retracted to full flap extension. If your plane is equipped with retractable gear, you should try the stalls with the gear extended and retracted. Also practice the stalls at different gross weights. You will find that the plane may react differently in each scenario, from the angle of attack at which the stall occurs, to the airspeed at which the stall occurs. As you stall the plane, be aware that the indicated airspeed may be inaccurate due to placement of the pitot tube and local airflow around it at high angles of attack. The operations manual for the plane should contain information related to correction

Fig. 2-16a Departure stall entry: forward view.

Fig. 2-16b Departure stall recovery: forward view.

Fig. 2-17a Departure stall entry: left side view.

Fig. 2-17b Departure stall recovery: left side view.

factors for the airspeed. This will allow you to determine the actual airspeed at which you are stalling and let you see how much stall speeds can differ based on the configuration of the plane.

Common Errors

I have seen a number of common errors during departure stall practice. The first is that students do not get the nose up steeply enough and the plane does not stall completely, but instead mushes along. While you do not want to rocket the nose skyward, use a smooth, firm pull on the control yoke to get the proper angle of attack to cause a stall. The next common error is neglecting to input sufficient rudder to keep the ball centered. In climbing turns to the right, some planes need a tremendous amount of right rudder to keep the ball centered. Unless you can "feel" the plane and know if it is coordinated, you should occasionally glance at the ball to judge if more or less rudder is required. If the plane is banking, some pilots let the angle of bank change, failing to hold a constant bank angle. In some cases, this can result in a very steep bank before the plane stalls. Finally, overcontrolling the amount of elevator necessary to recover is also a frequent error that pilots commit as they dive the airplane after the stall.

Slipping Departure Stall

This section covers the manner in which a Cessna 152 behaved in a slipping departure stall. In this stall sequence, the plane was configured in a standard departure mode, with the engine at full power and the airspeed at V_y. The plane was banked to the left approximately 30° in a coordinated, climbing turn. Initially airspeed was held at V_y. After the turn stabilized full right rudder was input, causing the plane to yaw to the right and the plane to enter a slipping turn to the left. At the same time the plane's nose was pulled higher by increasing elevator backpressure, with the airspeed soon bleeding off toward stall speed. As the plane stalled, the nose yawed to the right, in the direction of the cross-controlled rudder. A spin to the right developed immediately after stall entry, the nose dropping sharply, as is characteristic for Cessna 152 during spin entry. Ailerons were neutralized to avoid flattening the spin. Spin entry and recovery procedures are covered in Chapter 3.

The sequences in Figures 2-18 and 2-19 again show the front and side view to the left of the stall and spin entry. As you can see, the slipping departure stall differs from the slipping approach stall in that the plane does enter a spin, as opposed to continuing to buffet in an uncoordinated turn to the left. Again, I hypothesize the difference is due to the fact that at the slower airspeeds in the approach stall, the elevator and rudder do not have the authority to force the plane into a spin; while in the departure stall, the slipstream generated by the propeller aids the controls in causing the plane to enter a spin.

Common Errors

The common errors covered in the normal departure stall section also apply to slipping departure stalls. Even though uncoordinated rudder is used for this

Fig. 2-18a Slipping departure stall entry: forward view.

Fig. 2-18b Slipping departure stall entry: forward view.

Fig. 2-18c Slipping departure stall spin entry: forward view.

Fig. 2-19a Slipping departure stall entry: left side view.

Fig. 2-19b Slipping departure stall: left side view.

Fig. 2-19c Slipping departure stall spin entry: left side view.

maneuver, the bank may not be held constant. Not recovering from the stall immediately by coordinating the rudder and releasing elevator backpressure smoothly can cause the plane to quickly enter a spin, as was documented in this section. Because most departure stalls take place close to the ground, avoiding the spin should be the major concern. Generally, recovery from a stall will take much less altitude than from a spin. The spin is best avoided by not stalling at all, but flying with the rudder in a coordinated manner will reduce your chances of entering a spin if you do find yourself in a true departure stall.

Skidding Departure Stall

In this scenario, the plane was once again put into a departure climb, the engine generating full power and the flaps retracted. From that point a shallow, coordinated climbing left turn was entered. Once the turn stabilized, full left rudder was applied, quickly yawing the nose to the left. Right aileron input became necessary to prevent the bank from increasing as full left rudder was applied, further increasing the amount of crossed controls. Elevator backpressure was increased and the airspeed began to fall off rapidly. As the plane stalled, the nose dropped and the plane entered a spin in the direction of rudder input, in this case to the left. Ailerons were put in the neutral position and spin recovery was initiated.

Figures 2-20 and 2-21 show the forward and side views during the stall and spin entry. Compared to the pitch angles found in the approach skidding turn,

Fig. 2-20a Skidding departure stall entry: forward view.

Fig. 2-20b Skidding departure stall spin: forward view.

Fig. 2-20c Skidding departure stall spin entry: forward view.

Fig. 2-21a Skidding departure stall entry: left side view.

Fig. 2-21b Skidding departure stall spin: left side view.

Fig. 2-21c Skidding departure stall spin entry: left side view.

you can see that the nose is higher as compared to the horizon due to the power generated with the engine at full throttle.

Common Errors

The same common errors covered for normal and slipping departure stalls apply to skidding departure stalls. Stall avoidance is the best way to stay out of the spin that may develop in this sequence, and coordinated use of rudder in the event of a stall will also be of benefit.

Altitude Loss During a Stall

In any stall situation, whether practice at altitude or an accidental stall during an approach to landing, recovery from the stall with a minimum of altitude loss can make the difference between a successful or unsuccessful recovery. When practicing, you should critique yourself each time, noting how many feet were lost during the stall, as well as your stall recovery technique.

By improving your technique, you can reduce the altitude loss to the least possible in which your plane can recover from a stall. The next few paragraphs discuss the importance of adding full power and the effects it can have on the plane. They also discuss establishing a positive rate of climb as quickly as possible after recovery from the stall. Finally, common errors that result in additional altitude loss after a stall are covered. You must know how to minimize altitude loss if you do get into a stall or spin. Chapter 8 reviews a number of

NTSB stall/spin accident reports. The intent is to increase your awareness of the seriousness of stall/spin accidents. The examples in Chapter 10 document that pilots who seemingly should never have put themselves into the situation did not react correctly and made a potentially salvageable situation worse.

Engine Power During Stall Recovery

To recover from a stall, the most important step you can take is to reduce the angle of attack to less than the critical angle of attack as quickly and smoothly as possible. You then want to accelerate to airspeeds that will allow you to stop the descent and establish a positive rate of climb. Because most stall accidents take place during takeoff or landing, when your airspeed is low, the use of full throttle can help increase the airspeed sufficiently to avoid a secondary stall as you level the airplane after the recovery and minimize altitude loss. The additional thrust from the engine helps prevent the need for excessive nose down pitch angles that may be necessary in a true low altitude stall situation, when impact with the ground may be imminent. While these seem like commonsense concepts, the fact that pilots crash as a result of low altitude stalls each year indicates that not everyone understands their importance.

When the plane stalls, reduce backpressure to lower the nose to just below the horizon. The amount of nose down attitude will vary from plane to plane, and flight to flight with the same plane, depending on how the plane is loaded. Many flight instructors today teach students to push the nose well below the horizon, telling them the extreme nose down attitude is necessary to recover from the stall. In my opinion, in most general aviation aircraft this extreme measure is not necessary and, in fact, increases the altitude loss during recovery. Experiment with the plane you fly, but you will probably find that when the plane stalls, only minor nose down attitudes are required for stall recovery.

As the nose is lowered for stall recovery, you should also add full power. If you are holding any elevator backpressure, the nose of the plane may have a tendency to rise as power is added. Be prepared for this and reduce the amount of nose up elevator as necessary to avoid the plane pitching up and entering another stall. You should always make the elevator control inputs smoothly, avoiding the tendency to slam the control yoke forward or back. As you will recall from the discussion about accelerated stalls earlier in the chapter, this may result in an accelerated stall even when you are well above stall speed. Once full throttle is added and sufficient airspeed has been obtained, smoothly ease up elevator in to arrest the descent, and then establish a climb if necessary.

Here is a note of importance: if you should stall during an approach, the nose up trim you have added to control the airspeed can work against you as you try to recover from the stall. As power is added and airspeed increases, the nose up trim will try to force the plane's nose higher, increasing the likeliness of another stall. Some pilots feel that by forcing the nose of the plane down, against the trim, they may brake something. Outside of mechanical deficiencies with a plane, this is not true. If you need to, override the nose up trim by pushing forward on the control yoke to achieve the proper attitude. As always, do not overcontrol and aggravate the situation. Make the plane do what it needs to do to fly correctly.

Climbs After Stall Recovery

The extent of the climb needed after a stall depends on the situation. I have flown planes that lose very little altitude during stall recovery when it is done correctly, while others may lose much greater amounts of altitude. In very aggravated stall situations, you may lose several hundred feet before recovery is successfully completed. If you are close to the ground, you will want to climb to regain altitude; while if you are practicing stalls, you may already have sufficient altitude and do not need to climb. Keep in mind, however, that the more you practice a procedure, the more second nature it becomes. For this reason, it may be a good idea to always establish a positive rate of climb after recovery, even when practicing. In this manner, if an emergency stall or spin recovery is ever necessary, you will be in the habit of climbing once you have recovered from the stall.

Avoid the temptation to pitch the nose up too quickly, or at too high an angle of attack. This action could easily set you up for another stall; and at altitudes that may already be very low, recovery may not be possible.

Common Errors

When pilots practice stall recovery during flight training, or their biennial flight review, there is usually little emphasis on recovery technique or loss of altitude during the stall. As a result, poor technique in stall recovery does not manifest itself to the pilot as excessive altitude loss. Depending on the instructor they are flying with, poor techniques may even be reinforced. I recently had a discussion with an NTSB representative. He told me of a flight instructor that taught his students to "shove the nose down pretty far, 'cause you need to do that to recover from a stall."

A student that is taught this stall recovery technique has no idea that they are being taught stall recovery incorrectly; but I have run into pilots that think the nose needs to be well below the horizon to recover. This is the most common error in my experience that results in excess altitude loss. Not adding throttle during stall recovery, or not bringing in full throttle, is another common error. Some flight schools do not want to put excessive wear on engines, and feel full throttle during stall recovery may be too hard on them. Students then learn to use only partial engine power during stall recovery, again increasing the altitude lost. If you have learned these poor techniques, find an instructor that will help you unlearn them. The right instructor can turn stall entry and recovery practice from a tension-filled flight to an enjoyable flight experience.

SUMMARY

Many students and rated pilots approach stalls as though they are putting their hands into a fish tank filled with piranhas. They feel it is only a matter of time until they get into a stall from which they will be unable to recover. With proper training, this is completely untrue; and, in fact, proper practice increases your level of flight safety and proficiency. This chapter has covered a number of topics that are related to stalls. It began with a definition of stalls and the factors that affect the stall speed of an airplane. Remember that a plane can stall at any airspeed; so do not assume that if you are in the airspeed indicator's green arc, you are safe from stalls.

The chapter also reviewed accelerated stalls and how abrupt control movements and steep banks can greatly increase the stall speed of the plane. Accelerated stalls are of particular importance when flying in the pattern. Avoid steep banked turns when at the slow speeds you fly in the pattern. There is less margin above the stall speed that will result from steep banks. Planning your turns in the pattern and compensating for the effects of wind will help you avoid the need for banks that are steeper than safety dictates.

A number of different stall situations were discussed, many of them typical stall scenarios that pilots may get themselves into during departure and approach. As you will recall, the majority of these stall examples ended with spin entry. The stall/spin cases from NTSB included in Chapter 8 provide graphic examples of how common these types of stalls unfortunately are. The overriding theme of each of the stalls covered is that if you fly the plane in a coordinated manner, with correct use of rudder to keep the ball centered, you can avoid spins. With proper stall recovery techniques, you can recover from a stall while losing very little altitude. Remember, the best way to avoid a spin is not to stall.

3

Introduction to Spins

THIS CHAPTER FOCUSES ON NORMAL SPINS. FOR A SURPRISINGLY large percentage of pilots, this is an area of flight that they know little about. Once again, every pilot should receive spin training as part of private pilot training. Furthermore, a review of spin entry and recovery should be included as part of the biennial flight review to help pilots maintain at least a basic exposure to spins on a regular basis. Statistics documenting the number of stall/spin accidents the NTSB investigates, and the high percentage of fatalities in those accidents, illustrate why pilots can benefit from proper spin training.

This chapter is not filled with the equations or the lift-over-drag graphs that are found in many spin texts. Instead, it deals with the physical actions that take place during the life of a spin, including how the plane's motions change, how the controls react, and what you feel while in a spin. Formulae and tables have their place, but they have been reviewed a great deal in other texts and this chapter focuses on other aspects that directly affect you when you fly.

A number of subjects related to understanding spins are reviewed in this chapter. The first section covers spin testing for aircraft certification as required by the FAA. Not all aircraft are certified for spins, and those that are may have limitations in terms of the types of spins they can execute. Federal Aviation Regulations (FARs) pertaining to aerobatic flight, in which spins can be included, are also mentioned. Spins are defined and the relationship between stalls and spins is further explored. If you don't stall, you can't spin. Spin entry, both planned and accidental, is discussed, as well as the importance of control coordination.

The various phases of a spin are reviewed in detail. Many pilots are unaware that a spin has different phases or that an aircraft can react differently to each phase. The spin recovery discussions in this chapter cover improper recovery techniques and altitude loss during a spin. Executing recovery steps in the wrong sequence can seriously affect the success of the recovery. Instrument behavior during the spin and emergency spin recovery techniques are the last topics covered.

A great deal of information is packed into this chapter, but it is fundamental to your understanding of spins. There are many misconceptions about spins and, while you do not need to become an aeronautical engineer to understand them, you do need to know what causes spins, how they progress, and how to correctly recover from them. Keep in mind that not all planes behave in the same manner during a spin. The examples in this chapter use a specific aircraft model; but even with different planes of the same model, depending how they are rigged, their weight and balance, and many other factors, you will find a difference in spin characteristics. The information provided here is only a baseline to which you should add your own experiences.

SPIN CERTIFICATION OF AIRCRAFT

The FAA has set specific regulations related to the certification of aircraft: Part 23 of the FARs (see the end of this section). Among these are spin testing requirements for normal, utility, and acrobatic (also known as aerobatic) aircraft. To summarize these regulations, aircraft in the normal category should not be spun. For this category, manufacturers are required to execute only a one-turn spin or a three-second spin, whichever takes longer. As explained later in this chapter, during the first turn of a spin a plane is in the incipient phase, from which it is much easier to recover than from a developed spin. During a spin test, therefore, planes in this category remain in the incipient phase, making the test pilot's job of recovery much easier.

During spin testing, the manufacturer may also install a parachute on the tail of the plane to aid in emergency spin recovery. In the event the plane has trouble recovering from the spin, the chute can be released to help stabilize the plane and recover from the spin. As far as the flying public is concerned, past one turn, a plane certified in the normal category has spin characteristics that are unpredictable and, for this reason, spins should be avoided.

If you spin a normal category plane or one placarded against spins you may find that the plane is unable to recover, no matter what control inputs you use. "The one-turn 'margin of safety' is essentially a check of the airplane's controllability in a delayed recovery from a stall" (*Flight Training Handbook*, p. 155). Even utility and acrobatic category planes may get into unrecoverable spins if they are not properly loaded, or the manufacturer's operations manual regarding spins is ignored.

The *Flight Training Handbook* also states that many pilots ignore the dangers of spinning a normal category plane. According to the manual, "increasing occurrences involving airplanes wherein spin restrictions are intentionally ignored by pilots, have been brought to the attention of the FAA" (p. 155). The fact that the plane was spin-tested during its certification, to the level required by Part 23, does not mean that it can be spun safely. If you choose to ignore this directive, you are entering the world of the test pilot, with all of the risks that includes.

In many instances, past one turn the plane enters a flat spin, from which it may not be recoverable (see Chapter 7). Regulations can be very dry reading; but the potential for accidents during spins justifies the inclusion of regulations associated with spin certification.

FAR 23.221 states:

a. Normal category. A single-engine, normal category airplane must be able to recover from a one-turn spin or a 3-second spin, whichever takes longer, in not more than one additional turn, with the controls used in the manner normally used for recovery. In addition,

   1. For both the flaps-retracted and flaps-extended conditions, the applicable airspeed limit and positive limit maneuvering load factor may not be exceeded;

   2. There may be no excessive back pressure during the spin or recovery; and

   3. It must be impossible to obtain uncontrollable spins with any use of controls.

For the flaps-extended condition, the flaps may be retracted during recovery.

b. Utility category. A utility category airplane must meet the requirements of paragraph (a) of this section or the requirements of paragraph (c) of this section.

c. Acrobatic category. An acrobatic category airplane must meet the following requirements:

   1. The airplane must recover from any point in a spin, in not more than one and one-half additional turns after normal recovery application of the controls. Prior to normal recovery application of the controls, the spin test must proceed for six turns or 3 seconds, whichever takes longer, with flaps retracted, and one turn or 3 seconds, whichever takes longer, with flaps extended. However, beyond 3 seconds, the spin may be discontinued when spiral characteristics appear with flaps retracted.

   2. For both the flaps-retracted and flaps-extended conditions, the applicable airspeed limit and positive limit maneuvering load factor may not be exceeded. For the flaps-extended condition, the flaps may be retracted during recovery, if a placard is installed prohibiting intentional spins with flaps extended.

   3 It must be impossible to obtain uncontrollable spins with any use of the controls.

FARS RELATED TO SPINS

There are a number of FARs you should be aware of whenever you practice spins. I have spoken to a number of people about spins and whether they are considered aerobatic maneuvers. Predictably, some claim that they are; while others feel that, when executed as part of the required training for a flight instructor rating, they are not. The best way to avoid problems with spins is to adhere to the FARs that outline the requirements for aerobatic flight. (Yes, I dodged stating whether spins are an aerobatic maneuver.)

You should read FAR sections 91.303 and 91.307 (see the end of this section). In a nutshell, they state that you may not execute aerobatic maneuvers over a congested area, town, or open air assembly of persons; within Class B, C, D, or E airspace; within four nautical miles of the center of a federal airway; below 1,500 feet above ground level (AGL); or when flight visibility is less than three miles. This is all pretty straightforward, but in the next paragraph the regulations become somewhat more open to interpretation.

When you fly aerobatics, which include any maneuvers with a bank of more than 60° or a nose up or down angle of greater than 30°, you are required to wear a parachute. This parachute must have been repacked within the time period set for that type of rig, which is typically 120 days unless it is composed of silk or other natural fibers, which must be repacked every 60 days.

Now, if spin training is given that is required by regulations for any certificate or rating by a certified flight instructor (CFI) or airline transport pilot (ATP), parachutes are not required. Therefore, pilots taking spin training from a CFI can state that they are doing so in preparation for a flight instructor rating. In that situation, are parachutes required for spin training? This particular issue has been much debated. In many cases, CFIs will not use parachutes when giving spin training. In an official letter presented at a CFI refresher clinic the FAA stated that spin training given by a CFI was acceptable without parachutes. FAR 91.307 (d)(2), which states that parachutes are not required for "spins and other flight maneuvers required by the regulations for any certificate or rating...," can be interpreted to mean that spin training given by a CFI, even for non-CFI flight instruction, can be done without a parachute. Properly done, in a well-maintained aircraft, this is not a problem. But the safety call is ultimately up to you. When in doubt, wear a parachute, if for no other reason than that it is the ultimate insurance policy.

Following are the FARs 91.303 and 91.307:

91.303 Aerobatic Flight

No person may operate an aircraft in aerobatic flight—

a. Over any congested area of a city, town or settlement;

b Over an open air assembly of persons;

c. Within the lateral boundaries of the surface areas of Class B, Class C, Class D, or Class E airspace designated for an airport;

d. Within 4 nautical miles of the center line of any Federal airway;

e. Below an altitude of 1,500 feet above the surface; or

f. When flight visibility is less than 3 statute miles.

For the purposes of this section, aerobatic flight means an intentional maneuver involving an abrupt change in an aircraft's attitude, an abnormal attitude, or abnormal acceleration, not necessary for normal flight.

91.307 Parachutes and Parachuting

a. No pilot of a civil aircraft may allow a parachute that is available for emergency use to be carried in that aircraft unless it is an approved type and—
   1. If a chair type (canopy in back), it has been packed by a certificated and appropriately rated parachute rigger within the preceding 120 days; or
   2. If any other type, it has been packed by a certificated and appropriately rated parachute rigger—
      i. Within the preceding 120 days, if its canopy, shrouds, and harness are composed exclusively of nylon, rayon, or other similar synthetic fiber or materials that are substantially resistant to damage from mold, mildew, or other fungi and other rotting agents propagated in a moist environment; or
      ii. Within the preceding 60 days, if any part of the parachute is composed of silk, pongee, or other natural fiber, or materials not specified in paragraph (a)(2)(i) of this section.
b. Except in an emergency, no pilot-in-command may allow, and no person may make, a parachute jump from an aircraft within the United States except in accordance with part 105.
c. Unless each occupant of the aircraft is wearing an approved parachute, no pilot of a civil aircraft carrying any person (other than a crewmember) may execute any intentional maneuver that exceeds—
   1. A bank of 60 degrees relative to the horizon; or
   2. A nose-up or nose-down attitude of 30 degrees relative to the horizon.
d. Paragraph (c) of this section does not apply to—
   1. Flight tests for pilot certification or rating; or
   2. Spins and other flight maneuvers required by the regulations for any certificate or rating when given by—
      i. A certificated flight instructor; or
      ii. An airline transport pilot instructing in accordance with 61.169 of this chapter.
e. For the purposes of this section, approved parachute means—
   1. A parachute manufactured under a type certificate or a technical standard order (c-23 series); or
   2 A personnel-carrying military parachute identified by an NAF, AAF, or AN drawing number, an AAF order number, or any other military designation or specification number.

DEFINING SPINS

Depending on the text you read, spins are described in different ways. To understand what a spin is, you must understand what causes a plane to spin. A complete review of spin factors is not within the scope of this book, but the basics related to spins are covered. The next few paragraphs cover the physical actions that take place as the plane enters a spin. A complex set of physical effects start as the plane stalls, and then enters the spin.

Stall/Spin Relationship

When a plane enters a spin, it may have one or both wings stalled. For one reason or another, one wing has a lower angle of attack than the other. The wing with the lower angle of attack is less stalled and produces more lift. This can be due to a number of reasons, such as how the plane is rigged or uncoordinated use of the rudder and ailerons during the stall. For these reasons, the plane begins to yaw in the direction of the more stalled wing, causing the plane to slip in the same direction it is now yawing.

The situation further compounds itself as the side of the fuselage, the vertical fin, and other vertical surfaces weathervane the plane into the relative wind (*Flight Training Handbook*, p. 156). Figure 3-1 depicts the different angles of attack of the wings as the plane enters a spin. As you can see, the lower wing has a higher angle of attack than the raised wing.

Rolling also begins to take place due to the difference in lift being produced by the wings. The lowered wing has a larger angle of attack than the higher wing and produces less lift, causing the higher wing to roll toward the lower wing. The yawing, rolling, and other aerodynamic forces such as drag and centrifugal force cause the plane to continue in a nose down, rolling, yawing maneuver, which is a spin. The pitch angles and rate of rotation will vary during the course of a spin in many planes. Each plane may also react differently in its spin characteristics.

Once a plane is in a spin, it remains there until recovery is initiated. In a plane designed for spins, multiple-turn spins are completely safe. But for other aircraft, passing the one-turn point in a spin and entering a developed spin can be a problem.

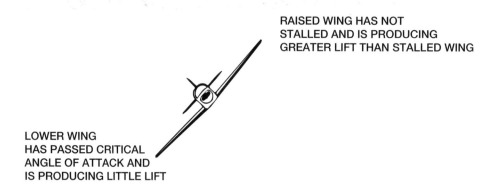

RAISED WING HAS NOT
STALLED AND IS PRODUCING
GREATER LIFT THAN STALLED WING

LOWER WING
HAS PASSED CRITICAL
ANGLE OF ATTACK AND
IS PRODUCING LITTLE LIFT

Fig. 3-1 Wing angle of attack differences at spin entry.

Spins combine a number of aerodynamic factors and are therefore a complex maneuver. Spins are also directly related to stalls because a plane must stall for it to enter a spin. This does not necessarily mean that your airspeed must be low; an accelerated stall can also result in a spin. In fact, the snap roll, which is an aerobatic maneuver, is essentially a high-speed stall that results in a spin. However, in the case of a snap roll, the spin may be horizontal—or any other direction, depending on how the maneuver was executed. Spins can even be performed going straight up!

Another point that should be considered relative to spin entry is that planes that are difficult to get into a spin, and must be forced into them, may also be difficult, if not impossible, to get out of a spin. The resistance a plane has to entering a spin is due to its inherent stability or other design factors. When the plane is forced into a spin through extraordinary means, such as excessively high angles of attack when the plane stalls, or keeping power in during an uncoordinated stall, the resistance it exhibits during entry may also be present when you attempt to recover. This is yet another reason to avoid spins in planes that are in the normal category or placarded against spins.

SPIN ENTRY

Many misconceptions surround entry into a spin. Some view it as a violent maneuver, with the plane out of control once the aircraft crosses over the line separating stalls from spins. Others are intimidated by their expectations of the unusual attitudes that they will encounter during the spin. Some pilots express concern over looking straight down at the ground during a spin. I have found that starting spin training with a very thorough preflight discussion of spins, covering the angles and sensations that the pilot can expect to experience during spin entry and recovery, helps reduce the fear of spins.

This section covers spin entry from two different view points: the consciously planned spin, such as you would execute during spin training or practice, and the unplanned spin entry. Unplanned entry is of greater concern, due to the fact that it is probably an accidental situation, often close to the ground during an approach or departure. However, do not take planned spin entry lightly; improperly executed, planned spins can get out of hand, as examples later in the book document.

Planned Spin Entry

Just as with stalls, the operations manual for the plane in which you practice spins should be the final word on spin entry and recovery procedures. The procedures discussed here are general in nature, but should give you an understanding of how to correctly enter a normal, upright spin. To begin, be certain you have climbed to a safe altitude in the practice area and performed clearing turns prior to setting up for spin entry. Spins usually result in greater loss of altitude than a stall, so give yourself enough extra altitude to recover from the spin safely. Just as with stalls, add carburetor heat as recommended by the airplane's manufacturer prior to reducing engine power. Spins should always be started in a power off configuration. For reasons discussed in Chapter 7, engine power above idle during spin entry can increase the chances of entering a flat spin.

Decide prior to entering the stall which direction you want to spin. Waiting until after the plane has wallowed along in a stall before making this decision, and then inputting the rudder, may cause the plane to take additional time and altitude before it begins to spin. As the plane stalls, ease in full nose up elevator, maintain neutral ailerons, and push full rudder in the direction you want to spin. For instance, if you want to do a spin to the right, push full right rudder. For spins to the left, use full left rudder. At this point, the plane will begin the rolling, yawing tendencies and enter a spin.

Unplanned Spin Entry

Unplanned spin entry can be a deadly situation. It may seem impossible to believe that pilots can put themselves into a situation from which a plane can start a spin accidentally. Given the number of low-altitude spin accidents that take place, it is much less difficult than you may suspect. Loss of engine power, incorrect weight and balance, improper go-around procedures, and uncoordinated low-altitude turns are only some examples of factors that precipitated stall/spin accidents between January 1992 and August 1995, as recorded by the NTSB.

The accident reports document that accidents often occur when the pilot is faced with a partial or full power loss during takeoff or landing. The pilot seems to quit flying the airplane, letting it get too slow or banked too steep, as he or she tries to force the plane back around to the airport. In these cases, picking an alternate landing site off airport may have resulted in a controlled forced landing instead of a loss of control of the plane in a steep, slow, low-altitude turn.

Low passes down the runway, followed by steep pull-ups, also cause a number of accidents. This can be due to a number of factors; but as the plane slows in the climbing turn, it is in a prime setup for a stall/spin. High angle of attack, low airspeed, and left yawing tendencies are all characteristics that position the pilot for a potential accident. Once again, the pilot gets behind the airplane and realizes too late that the plane has gotten out of control.

Steep banked turns, resulting in accelerated stalls and spins at low altitude, are also a problem. Strong winds, and poor compensation for their effects during turns to final, have also caused planes to enter stall/spin accidents. In these cases, the pilot attempts to force the plane to the runway centerline during the turn to final approach.

Accidents recorded by the NTSB are covered in more detail in Chapter 8, but this brief discussion indicates the types of situations that can result in an unplanned entry into a spin. Your first priority is to fly the plane. If you let it fly itself, or attempt to force it to fly outside its performance envelope, you are setting yourself up for a potentially serious accident. Planning ahead, knowing the environment you are flying in, and the performance characteristics of the plane you are flying can help you end every flight safely.

Importance of Control Coordination

Control coordination is the key to avoiding spins. The approach and departure stalls reviewed in Chapter 2 are graphic testimony to the fact that if a plane stalls and is slipping or skidding, it has a strong tendency to enter a spin. The moral of the story should be clear. When you fly, use the rudder correctly and keep the ball centered to reduce the possibility that you will enter a spin.

You should get in the habit of using the correct amount of rudder in all phases of flight, and not just during slow flight. Right rudder during climbs and rudder during turns should be an automatic response. Many pilots fail to realize that they may need left rudder during descents to keep the ball centered. Just as the descending blade of the propeller has a greater angle of attack during a climb and generates more thrust, the ascending blade during a descent has a greater angle of attack than the descending blade. As a result, the plane has a tendency to yaw to the right and left rudder becomes necessary to maintain coordinated flight. The amount of rudder necessary varies from plane to plane, so experiment with the plane you fly to see how much you need to keep the ball centered during a descent with power off and with power on.

CONTROL POSITIONS DURING A SPIN

The next series of paragraphs cover how controls should be positioned in a normal, upright spin. Improper use of controls during spin entry, while in the spin, or during the recovery can dramatically affect how a spin develops. In addition to reviewing how the elevators, rudder, and ailerons can affect a spin, the discussion covers the relationship between the elevator and rudder. The elevator's positioning and the design of the rudder can determine how effective the rudder is during spin recovery.

Spin Entry

Normal spins are executed from straightahead, power off stalls. As the plane stalls, full rudder is input in the direction you want to spin. To spin left, full left rudder is used; to spin to the right, full right rudder is used. At the same time rudder is input, the elevator control is moved to the full aft position and the ailerons are kept in the neutral position.

Not using full rudder, or not maintaining full up elevator as the plane stalls, can cause the plane to take longer to enter the spin, mushing along in an uncoordinated stall and losing altitude. Many students use partial rudder input when first learning spins. The plane yaws in the direction of the rudder and eventually enters either a spiral or spin. When you input rudder, it should be a firm, brisk but not violent, movement of the rudder pedal. This will aid the plane in making a clean entry into the spin.

The same is true of elevator controls as you enter the spin. Depending on the plane, how it is loaded, and other factors, you may not need full aft elevator input to cause the plane to stall. This is fine for stall practice; but when you enter a spin, as the plane stalls and you input rudder, you should move the elevator control to the full aft position. This helps ensure that the plane remains in a stall during spin entry and again aids in a good, clean start to the spin.

As with the rudder, students are often hesitant to input full up elevator, not being sure of the plane's response. In addition, given the rapid pace of events as the plane moves into the spin, students have been overwhelmed and forgotten to move the elevator control to the full aft position. Each plane will respond differently to only partial use of elevator during the spin; but this can also cause the plane to wallow along before entering a spin, or prevent the plane from entering a spin at all. Sometimes the plane recovers from the stall as a result of the change in the angle of attack brought on by the plane's nose dropping as rudder was applied.

Ailerons should be maintained in the neutral position during your entry into a spin. The terms "pro-spin" and "counter-spin" will be used through the rest of the book. Input of pro-spin ailerons, or ailerons that cause the plane to bank in the same direction as the spin, can result in a higher rate of rotation during the spin or extreme bank angles. Pilots who are in the habit of keeping ailerons and rudder coordinated tend to enter aileron in the same direction as the rudder input during spin entry.

A relatively large percentage of students, when first exposed to spin entry, attempt to input pro-spin aileron as they input rudder and bring the stick back to the full aft position. As noted in Chapter 2, improper use of aileron can cause a more complete stall of one wing, resulting in a greater lift differential between the wings and therefore a potentially higher rotational rate. In the aerobatic plane I fly, input of pro-spin ailerons once in a spin puts the plane into a knife-edge spin, with the plane spinning on a wingtip, almost 90 degrees to the horizon. This is fun in a plane that is capable of executing it, but not something you want to execute in a Cessna 152.

Use of counter-spin ailerons, or ailerons opposite the direction of spin rotation, can affect the spin in a potentially serious manner. For instance, in a spin to the left, inputting right aileron has the effect of raising the inside wing and the plane then flattens its rotational mode. Flat spins are covered in detail in Chapter 7, but counter-spin aileron input is one of the steps taken in entering an upright flat spin. Again, in the proper plane these types of spins are no problem. In the wrong airplane, use of counter-spin ailerons can result in a very serious situation. The best way to avoid problems is to keep the ailerons neutral as you enter the spin.

During Developed Spin

Once the plane has entered a spin, you need to maintain proper use of controls to keep the plane spinning in the correct manner. When pilots first experience spins during training, they are often surprised at the plane's reactions and the sensations they are experiencing. This can result in an unconscious relaxation of rudder and elevator inputs as the plane moves through the first one-half turn of the spin. A later section of this chapter discusses the various phases of spins; suffice it to say that during the first two turns or so of a spin, the plane has not really entered into a full-fledged spin. When this unconscious relaxation of elevator and rudder inputs takes place, the plane may recover from the spin and enter a spiral dive.

Given that most flying is done with much less than full elevator and rudder input, pilots naturally hesitate to use full control deflection during spin training. They have spent a flying career learning how to be smooth, making small control inputs and avoiding overcontrolling the plane. These are all good flying techniques; but to properly execute spins, you must adapt your control inputs to the requirement of the maneuver. Be sure to keep full aft elevator and full pro-spin rudder input during the entire spin. As during spin entry, ailerons should be maintained in the neutral position in order to keep the plane from increasing its rate of rotation, or going into a flat spin mode, during the developed portion of the spin.

Spin Recovery

When pilots are beginning spin training, learning the use of controls during recovery from the spin is the highest priority they seem to have, with good reason. As a pilot for skydivers, I was once told, "all the other pilots have made a jump, now you need to." During the freefall from 10,500 feet to 4,000 feet, the overriding thought in my mind was, "I sure hope this chute opens." To my great relief, it did, even though it took me almost two days to get over the adrenaline rush. I equate the focus I had on the chute opening that day to the feeling that first-time spin students are experiencing, wanting to be sure that they understand how to get out of the maneuver they are entering.

The operations manual for the plane you fly should be the final word in spin recovery procedures, but correct use of controls in the proper sequence is the key to smooth, controlled spin recovery. To begin, rudder opposite the direction of the spin, or counter-spin rudder, should be applied. This has the effect of slowing the rate of rotation. Next, apply a positive, forward elevator movement. This will vary depending on the plane, but it may be necessary to make this a brisk movement to get the plane to properly recover from the spin by reducing the angle of attack, thus breaking the stall. Once the plane stops spinning, neutralize the rudder and return to level flight (*Flight Training Handbook*, p. 157). Figure 3-2 illustrates the actions a pilot should take according to the FAA.

During spin recovery using the controls properly becomes even more important. Use of only partial counter-spin rudder can result in the plane taking longer to recover from the spin or not recovering at all. As with spin entry, input of rudder opposite the spin direction should be smooth, firm, and brisk, but not violent. Use full rudder travel, not just a partial input of opposite rudder. I have had numerous situations during which students did not use full

CLOSE THROTTLE
FULL OPPOSITE RUDDER
BRISK FORWARD ELEVATOR

HOLD ELEVATOR FORWARD
NEUTRALIZE RUDDER

EASE ELEVATOR BACK
TOWARD NEUTRAL

Fig. 3-2 Spin recovery procedures.

opposite rudder and the plane kept right on spinning, the student confused by the plane's apparent lack of response to the control inputs. After they are reminded to use full rudder input, the plane finally stops its rotation.

The elevator input should be brisk; but like during stall recovery, it does not need to pin you against the seat belt in a negative g maneuver. This, in fact, is a common reaction of students during spin recovery. The nose down attitude for spin recovery is normally greater than stall recovery; but I have had students jam the stick forward, pointing the nose straight down at the ground while any dirt on the floor floats up to the top of the canopy. All planes are different, but a forward movement of the elevator control does not normally need to be quite this enthusiastic. As during spin entry and the spin itself, ailerons should be kept in the neutral position. After the controls are coordinated, the plane should then be smoothly flown back to straight and level flight. Here again, beginning spin students are often very reactive in the use of controls to return to level flight, pulling back harder on the elevator control than is generally necessary.

RUDDER/ELEVATOR EFFECTIVENESS

Placement of the elevators relative to the rudder can have direct effect on the authority the rudder has during spin recovery. The illustrations in Figures 3-3a and 3-3b show an elevator in two different locations. In Figure 3-3a, the elevator is located at approximately the middle of the rudder, while the elevator in Figure 3-3b is located near the bottom portion of the rudder. During a spin the relative wind is at such an angle that the elevators can blanket airflow over the rudder, reducing the amount of authority it has in stopping the rotation of the spin. In Figure 3-3a, the lower portion of the rudder is not as completely blanketed by the elevators as the rudder in Figure 3-3b, the darkened area illustrating the blanketed area. This allows greater airflow over the rudder in Figure 3-3a and increases the chances of recovering the plane from the spin.

If you are contemplating spins in an airplane with the elevator located near the lower portion of the rudder, be certain it is approved for spins. Aircraft manufactured under part 23 of the FARs will document whether the plane is certified for spins. Experimental aircraft are another matter, though. I have flown experimental aircraft that were works of art, so the comments that follow are not a general judgment about those aircraft, but rather about the need to know the plane you are flying.

The operations manual may be incomplete for experimental planes—little more than pamphlets in some cases. Placards may be missing as well. Crucial information related to the ability of the plane to spin may not be readily available. Noticing the placement of the elevator can help you at least guess as to the amount of rudder that will be blanketed in a spin. You should never spin a plane until you know that it is safe to do so, though. Calling the manufacturer to confirm the ability of the plane to spin should be done before you fly the maneuver in an experimental airplane.

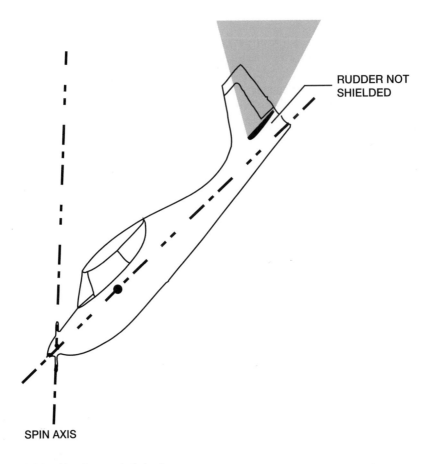

RUDDER NOT
SHIELDED

SPIN AXIS

Fig. 3-3a Mid rudder placement of elevator.

If you ever get a close look at high-performance aerobatic planes, notice how large the rudder is compared to other general aviation planes. Figure 3-4 shows the rudder of a Pitts Special. It is quite large compared to those found on planes of similar size. This additional rudder provides the degree of control the plane needs not only during spins but during aerobatic maneuvering. Because the plane may be tumbling through the sky, on purpose in most cases, the larger rudder area not only gives the pilot additional authority to recover from spins but helps overcome any airflow blanketing that may result from other unusual attitudes.

SPINS VERSUS SPIRALS

New spin students often enter a steep spiral instead of a true spin when they first begin practicing spins. As noted earlier in the chapter, not using full rudder or up elevator during spin entry can cause the plane to unintentionally

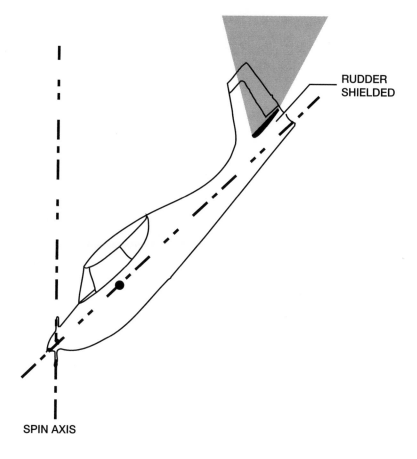

RUDDER
SHIELDED

SPIN AXIS

Fig. 3-3b Low rudder placement of elevator.

recover from the spin soon after it begins, or not enter a spin at all. In some cases, the plane may enter an attitude that resembles a spin, while it is actually spiraling down in a tight turn. Student pilots unaware of the difference, may still believe they are in a spin, because many characteristics that they attribute to a spin are present. There is a very real difference between the two maneuvers; and attempting to recover from a spiral using spin recovery techniques can make the situation worse, not better. This section examines the differences between the two maneuvers, and how to recognize when you are in a spin or a spiral.

By now you should have a good understanding of a spin: the plane has stalled, an asymmetrical lift situation is present, and the plane is in uncoordinated flight, yawing and rolling in a nose down attitude (at least for normal, upright spins). As you sit in the cockpit, the plane is nose down and turning rapidly in the direction of the spin.

Fig. 3-4 Pitts Special rudder.

Compare this "pilot's eye view" of a spin to the view of a spiral from the same perspective. In a spiral, the plane is in a nose down attitude, turning rapidly, bearing a strong similarity to the spin. What then, is the difference between the two maneuvers? To begin, when a plane is in a spiral, it is not stalled. Instead, it is rapidly gaining airspeed as the plane dives, while in a spin the airspeed hovers near stall speed. I have done multiple turn spins in the Pitts S-2B I fly; and throughout a spin, the airspeed remains constant, not increasing as the spin continues. This is the first, and key, difference between spins and spirals. If the airspeed is increasing, you are not in a spin. Figure 3-5 illustrates the difference in the behavior exhibited by the airspeed indicator. Figure 3-5a shows how the airspeed indicator stays near stall speed while in a spin, while 3-5b shows the airspeed indicator increasing as the plane accelerates in a spiral.

Attempting to recover from a spiral using spin recovery techniques can make the situation worse. In a spiral, the plane's airspeed may rapidly climb toward redline. Forcing the nose down in the mistaken belief that this will reduce the angle of attack and break the stall will only increase the airspeed more rapidly. In a spiral, you want to reduce engine power and gently ease the nose back to level flight. Ailerons will be necessary to roll the wings to level. Reducing engine power will help keep the airplane from accelerating too rapidly. Easing gently back on the elevator controls will help avoid an accelerated stall, or overstressing the airplane's structure. By contrast, adding up elevator in a spin, mistakenly believing you are in a spiral, will only help keep the airplane in the spin. Using ailerons during the spin to stop the rotation instead of rudder may cause the spin to go into a flat mode.

Fig. 3-5a Airspeed indicator in spin.

Fig. 3-5b Airspeed indicator in spiral.

You must understand the difference between spins and spirals and the correct procedures for recovery from each of them. Until you have experienced each of them, and understand their effect on the "feel" of the plane, you may need to rely on the airspeed indicator to differentiate between them. During your flight instruction in spins, be sure to have the instructor demonstrate the difference between these maneuvers.

SPIN PHASES

While in initial spin training, students often find that during the spin there are so many sensory inputs that they are overwhelmed by what is taking place. Later, after achieving a higher degree of comfort, they begin to understand and recognize the different phases that a plane passes through during a spin. You will come across different names for these phases; in this book, the four spin phases are called the spin entry phase, the incipient spin phase, the developed spin phase, and the spin recovery phase.

Spin Entry Phase

Spin entry has been touched on several times previously in this chapter. Spin entry is started from a normal, power off stall after performing clearing turns in the practice area. Carburetor heat should be applied per the manufacturer's recommendations. As the plane stalls, full elevator backpressure and rudder in the desired direction of the spin should be input in a smooth, brisk motion. This will cause the plane to transition from normal flight to the second phase of a spin, the incipient spin.

Incipient Phase

The incipient phase of a spin takes place as the plane is transitioning from forward flight to the nose down, rolling, yawing descent found in the spin. As the plane moves from forward flight into this attitude, it is affected by a number of factors, including inertia, lift, drag, and yaw. NASA has spent years studying spins, and entire aeronautical engineering texts could be devoted to the forces acting on a plane during the incipient phase. This section is limited to a high-level discussion of just a few forces.

As the plane enters the spin, inertia attempts to keep it moving forward in the direction of the original motion. This inertia must be overcome during the incipient phase of the spin to change the direction of motion of the plane. Like inertia, other forces such as lift, yaw, and drag must also transition to the new direction of motion for the plane. This transition takes time to complete.

Most single-engine, general aviation aircraft take approximately two turns to move from the incipient phase to the developed phase of the spin. In general, during the incipient phase of the spin it is much easier to recover from the spin, with the plane still not completely settled into the motions of a developed spin. While not an approved spin recovery technique, some single-engine, general aviation aircraft will recover from the incipient phase of a spin if the pilot merely releases the controls. DO NOT assume this is the case for the plane you are flying and go off to practice spin recovery on your own. For a host of reasons, this technique may not work and self-taught spin training is not the time to find out that the plane you are flying is the exception to this rule. At the end of the incipient phase, the plane has transitioned into a fully developed spin and it is much less influenced by its original movements.

Developed Spin Phase

As the plane moves from the incipient phase into the developed phase of the spin, it will settle down into a repetitive cycle. In the Pitts S-2Bs I fly, as the plane enters a spin, the nose drops to the lowest pitch angle at the 180° point of rotation, then begins to rise again, reaching its highest point at the 360° point. In the following turns, this pattern continues. The amount the nose rises and falls varies based on how the nose of the plane was as it stalled, how rapidly rudder and elevator backpressure were added, and other factors. A Cessna 152 also exhibits a similar repetitive pattern, though the nose down angles seem to be more extreme than for the Pitts Special.

Like the pitch angle during the spin, the rotational rate of the plane can also vary during the spin, speeding up and slowing down as the spin progresses. Each plane can vary in how much the pitch and rotational rate changes during the spin. In fact, the same plane will react differently from one spin to another, depending on factors at spin entry such as pitch angles and how the controls were applied.

I have had students execute stalls from very high nose up angles, followed by spin entry. In these stall/spin situations, the plane's rate of rotation and pitch changes have been different than when shallower pitch angles were used during spin entry. As you learn spins, try to pick up the sometimes subtle changes the plane goes through during the spin.

Spin Recovery Phase

Spin recovery has been covered to some depth earlier in this chapter, and further detail on specific spins is provided in later chapters. The spin recovery phase takes place as the pilot recovers from the spin, stopping the rotation the plane has entered, and breaking the stall. Like the incipient phase of the spin, planes may take what seems like a disconcertingly long time during the recovery to stop the rotation. Pilots, accustomed to the almost immediate recovery possible during the incipient phase, become unsettled if the spin continues after they have input the proper recovery controls.

According to the FAA (*Flight Training Handbook*, p. 157), the proper spin recovery technique is as follows:
1. Retard power.
2. Apply opposite rudder to slow rotation.
3. Apply positive forward elevator movement to break stall.

4. Neutralize rudder as spinning stops.

5. Return to level flight.

Some planes may take several turns after recovery control inputs have been completed before they stop the rotation. If a plane does not seem to be responding to their recovery inputs, pilots occasionally begin to try other control inputs. This approach only increases the amount of time they spend in the spin and the amount of altitude lost. If the recovery procedure in the operations manual of your plane is different from the one just outlined, use the manual. Be prepared for recovery to take several turns to complete.

IMPROPER RECOVERY PROCEDURE RESULTS

Use of improper spin recovery procedures can cause a plane to react in a variety of ways, few of them desirable. Depending on the way in which recovery is initiated, these responses can range from a few additional seconds of adrenaline pumping into your system, to damage to the plane, or, finally, to an impact with the ground. Many of the pilots I have worked with do not realize that the sequence of steps taken during spin recovery are as important as the steps themselves.

When you execute spins, always follow the sequence of recovery steps the aircraft manufacturer recommends. No text can cover all errors that are possible during recovery from spins, nor can it predict how all aircraft will react. Keep in mind that the plane you are flying may react differently from the examples provided here, perhaps because of control effectiveness or weight and balance. This admonition applies to such areas as elevator use, rudder use, and aileron use, all of which are discussed in the following sections.

Before considering possible errors, take another look at the standard spin recovery techniques as defined by the FAA (*Flight Training Handbook*, p. 157). The text accompanying this list in the *Flight Training Handbook* states: "To recover from the spin, the pilot should first apply full opposite rudder; then after the rotation slows, apply brisk, positive straightforward movement of the elevator control (forward of the neutral position). The control should be held firmly in this position. The forceful movement of the elevator will decrease the excessive angle of attack and thus will break the stall. When the stall is broken the spinning will stop. This straight forward position should be maintained and as the spin rotation stops, the rudder should be neutralized" (*Flight Training Handbook*, p. 157).

The text further documents: "Slow and overly-cautious control movements during spin recovery must be avoided. In certain cases it has been found that such movements result in the airplane continuing to spin indefinitely, even with the application of full opposite controls. Brisk and positive operation, on the other hand, results in a more positive recovery." Finally, "after the spin rotation stops and the rudder has been neutralized, the pilot should begin applying back elevator pressure to raise the nose to level flight. Caution must be used so as not to apply excessive back pressure after the rotation stops. . . . To do so will cause a secondary stall and may result in another spin, more violent than the first" (*Flight Training Handbook*, p. 157).

As you will see, taking these steps in the wrong sequence can result in significantly worse spins, rather than spin recovery. Let's begin with the use of the throttle.

Effect of Throttle on Spins

As a plane begins to enter a spin, you should always retard the throttle to idle immediately, if you did not stall in a power off configuration. As the plane begins to rotate, having power applied can cause the plane to raise the nose and cause the rotation to flatten out. Figure 3-6 shows how the plane transitions from the normal nose down spin attitude to a flatter spin attitude as a result of the thrust being generated by the engine. Recall that most single-engine airplanes are designed so that when power is applied, the thrust line is above the center of gravity and will cause the nose to rise. As explained in Chapter 7 one of the major steps in entering a flat spin, whether upright or inverted, is to apply engine power.

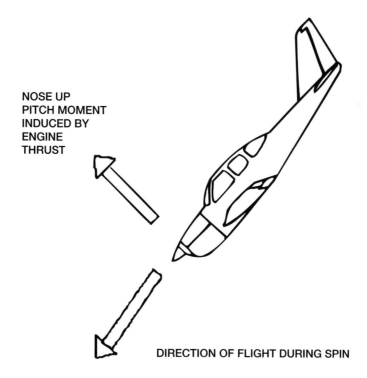

NOSE UP PITCH MOMENT INDUCED BY ENGINE THRUST

DIRECTION OF FLIGHT DURING SPIN

Fig. 3-6 Pitch change due to engine thrust.

I have taught many aerobatic students a maneuver known as the hammer-head. In this maneuver, the plane climbs vertically while generating full power. As it comes to almost a standstill, right, forward control stick is applied while left rudder is input. If properly executed, the plane pivots, and then flies a vertical downline before pulling back to straight and level. Done incorrectly, the cross-controlling combined with very slow airspeeds and a high power setting have resulted in students entering inverted flat spins. The first step taken to recover from the flat spin is to reduce power to idle immediately. This helps get the nose oriented to pitch angles that will allow more successful spin recovery to take place.

Once you have recovered from the spin, be sure to add power, but do so in a controlled manner. Especially for fixed pitch props, if you are traveling at high airspeeds as a result of a dive after the spin, it is possible to overspeed the engine by advancing the throttle too rapidly.

Effect of Rudder on Spins

The rudder plays a crucial role during spin entry and recovery. It is a common error for students to input only partial rudder during spin recovery. While in the incipient phase of the spin, this may work to slow the rotation of the spin enough to recover from the spin. However, in a fully developed spin, partial input of rudder opposite the spin may do nothing to slow or stop the spin rotation.

I once had an aerobatic student execute an inverted spin. Just at the plane was about to stall, the individual pushed forward on the stick (remember, we were upside down), raising the nose well above the horizon as rudder was added to enter the spin. The nose dropped to a very steep nose down attitude as the spin took hold. Then, forward pressure was relaxed on the stick, allowing the plane to enter an inverted, accelerated spin (for more detail on these spins, see Chapter 5).

At this point, we had a real ride going, the plane pulling just over two negative g's, rapidly rotating in an accelerated spin. The student applied partial opposite rudder, but the plane kept right on spinning. I then pushed full opposite rudder, and the rotation stopped. This was a very graphic example to this student that FULL rudder is necessary to recover from a developed spin.

Once the spin has stopped, not neutralizing the rudder can result in the plane yawing in the direction of the applied rudder. In severe cases, the plane may enter yet another spin, in the other direction. I have students practice executing this maneuver to help them understand the importance of neutralizing the rudder after the initial recovery. In some cases we have gone from the original spin, to a spin in the opposite direction, then back to a spin in the original direction, just by using rudder. You must apply the correct rudder, and then neutralize it as the rotation stops.

Effect of Elevator on Spins

Elevators play a crucial role in spin entry and recovery. As noted elsewhere, not having enough backpressure on the elevator control can result in an incomplete stall, and the plane will not properly enter a spin. Once the plane has entered a spin, incorrect use of the elevator can also be a problem during spin recovery. Not using sufficient forward elevator force can prevent the plane from recovering from the stall properly. Using too much forward elevator can cause the plane to dive steeply as you recover from the spin, thus losing excessive altitude. Remember that the elevator does not recover the plane from a spin; the rudder is used to stop the spin rotation. The use of forward elevator reduces the angle of attack and breaks the stall, which is what began the entire spin sequence!

Unless the aircraft's manufacturer indicates otherwise, you should input forward elevator only AFTER you have input rudder opposite the spin. Many pilots are under the impression that all they need to do is get the right control inputs in, without regard to the order in which they are performed. Accelerated spins show the pitfalls of this approach. An accelerated spin is the result of a change in the angle of attack during the spin and causes an increase in the spin's rate of rotation. Accelerated spins are entered by pushing forward on the stick while still holding pro-spin rudder input. As the control yoke is eased forward, the plane can dramatically increase its rate of rotation. The first time pilots experience an accelerated spin, they understand the importance of providing control inputs in the appropriate order.

To make a bad situation worse, once an accelerated spin begins, entering counter-spin rudder can cause the airplane to slip into an inverted spin. This can take place without an unsuspecting pilot realizing the plane has gone from an upright to an inverted spin. I attended an aerobatic safety seminar during which a pilot described how he had given the wrong recovery control inputs during an upright flat spin, and then slipped into an inverted flat spin. It took several turns for him to realize what had happened.

Once the rotation has stopped and the stall is broken, you need to fly the plane back to straight and level flight. Do not over control the elevator during recovery. Excessive backpressure can cause an accelerated stall and potentially cause the plane to enter another spin. Depending on your airspeed during recovery, you may overstress the plane's structure if the pull out results in excessive g forces.

Effect of Aileron on Spins

Ailerons should be maintained in the neutral position during a spin. I have had students use both counter-spin ailerons, which are ailerons rolling the plane in the direction opposite the spin, and pro-spin ailerons, which bank the plane even further in the same direction as the spin rotation. In either case, the plane reacted differently than the student expected.

In the case of counter-spin aileron input, a plane will have a tendency to flatten its spin rotation. Studies have found that planes may be more difficult to recover from a spin in a flat mode, in which the nose of the plane is not pitched down enough. In some cases, a plane in a flat mode may not be able to recover at all.

With pro-spin aileron, the plane will increase its angle of bank during the spin. Once in a spin, I add pro-spin aileron. The plane then banks up to almost a knife-edge flight, the wings at a 90° angle to the horizon. At this point, the plane seems to be spinning around the lowered wing tip. You do not want to execute this type of spin in a normal airplane, given the unusual loadings it can place on the plane.

Effect of Flaps on Spins

Flaps can affect the pitch angle that a spinning airplane will achieve. Airplanes equipped with flaps may be placarded against spins with flaps extended. While not conclusive, tests I have performed have shown that a plane will spin in a flatter mode with the flaps extended than when they are retracted. Depending on the aircraft, you may have to retract the flaps to help put the plane in a nose down attitude in order to recover from a spin. If a plane you are spinning begins to enter a flat mode, recover immediately. The primary indicator that a plane is entering a flat mode is the tendency of the nose to rise as the spin progresses.

Diving After Spin Recovery

After students stop a spin, many are struck by the nose down attitude of the plane. Figure 3-7 shows the view over the nose from a Cessna 152 just after recovery from a spin. As you can see, the nose is well down and a great deal of real estate is visible through the windshield. In some cases, pilots are not quite certain how to react, and do nothing. In these situations, the plane can quickly build airspeed, rapidly passing into the yellow arc on the airspeed indicator, toward redline. Be prepared for the nose down attitude you may experience after the spin, and recover smoothly back to straight and level.

This recovery to level flight should be a firm, smooth pullout. If the airspeed is climbing rapidly as you recover, you are probably not using enough elevator backpressure to raise the nose of the plane. One mistake of spin students is to use excessive forward pressure on the control stick to break the stall after rotation stops. In these cases, the plane will often end up on a vertical downline, the plane pointed straight down. This can rapidly build unnecessary airspeed and can be avoided by not using excessive forward elevator input.

Secondary Stall/Spin During Spin Recovery

At the other end of the spectrum from diving after spin recovery, some students use far too much nose up elevator backpressure during spin recovery. This can cause the plane to enter a secondary stall instead of recovering from the spin. If you fail to neutralize the counter-spin rudder input after recovery, the plane can enter a secondary spin in the direction of the rudder.

To avoid this situation, use a firm, smooth input of elevator backpressure during the recovery. If the plane begins to rumble, you are probably getting a

Fig. 3-7 Spin recovery: forward view.

high-speed buffet and are using too much backpressure. Another sign that you may be pulling too hard on the elevator is the feeling of excessive g forces. If you are being pressed down into your seat hard, you are likely attempting to force the nose of the plane up too quickly. With a little practice, you will know the correct amount of elevator backpressure to use to thread the needle between diving too steeply and pulling up too quickly as you fly the plane back to level flight.

Slow Spin Recovery

The final improper spin recovery technique covered here is slow response from the pilot in recovering from a spin. This is often only a matter of practice. Some pilots will fall behind in executing spin recovery procedures, increasing the amount of altitude lost during a spin. In a true, low-altitude, accidental spin situation, recovering from the spin in the minimum time possible may very well make the difference. For this reason, pilots should consider practicing spins on a regular basis with a qualified flight instructor. New spin students are often so caught up in sensations associated with a spin, they can take an inordinate amount of time before they react. This can result in a tremendous amount of altitude loss during the beginning stages of a spin.

Students generally come up to speed after practicing spins several times during a lesson, but then they must sometimes re-learn how to react quickly during the next lesson. Taking spin training only once will help you learn the mechanics of spin recovery, but you need more experience and repetition to react quickly enough to recover in a low-altitude situation. For this reason, practicing spins on a regular basis is necessary to be truly competent in rapid recovery from an emergency spin.

ALTITUDE LOSS DURING SPINS

Pilots are often astounded at the amount of altitude they lose during a one-turn spin. The amount of altitude lost varies from plane to plane and depends on the techniques used by the pilot; but in most cases, pilots lose approximately 1,000 feet of altitude during a one-turn spin. This fact should make you think very hard about spins entered from the pattern. A spin entered while on downwind MAY be recoverable; a spin entered on a turn from base to final is much more challenging to recover from safely. Spin avoidance is the best insurance policy while in the pattern, making stall recognition and coordinated use of the controls key components to never entering a spin.

Much of the altitude loss in a one-turn spin is the result of initial spin entry, as well as the altitude lost after the spin rotation has stopped, while flying the plane back to level flight. Spins with more than one turn show less altitude loss per spin, suggesting that there may be only a few hundred feet lost in a one-turn spin. Figure 3-8 is a table showing the number of feet lost in a one-turn spin, a three-turn spin, and a five-turn spin. As you can see, the average altitude lost per turn decreases as the number of turns increases. Use of proper spin recovery techniques can reduce the amount of altitude lost during a spin, and regular spin recovery practice can also decrease the altitude loss. The best plan is not to get into an unexpected spin, especially at low altitude.

INSTRUMENT BEHAVIOR DURING A SPIN

While you are in a spin, the instruments can behave in a seemingly illogical manner. In particular, gyroscopic instruments will behave in an unpredictable fashion, and the airspeed indicator also may seem to act irregularly. By understanding how your instruments may react, and what references you should use during spin recovery, you can increase your chances of recovering from a spin as rapidly as possible.

NUMBER OF TURNS	CESSNA 152	PITTS SPECIAL S-2B
1	600 ft.	800 ft.
2	1,100 ft.	1,500 ft.
3	1,500 ft.	2,000 ft.

Fig. 3-8 Spin altitude loss table.

Gyro Instruments

Flight instruments that use a gyroscope, whether vacuum or electrically driven, will behave in an erratic manner during a spin. Due to the unpredictability of these instruments during a spin, you should not rely on them to indicate the direction of the spin or your attitude during the spin.

During a spin, gyroscopes will initially attempt to hold in a fixed attitude. Once the plane's attitude passes a certain point, the gyroscopes hit the stops of the mounts that hold them. When this happens, they lose their orientation and begin to tumble. Gyro-driven instruments, such as the artificial horizon and directional gyro, then indicate unreliable information because the gyroscopes are unable to maintain their orientation.

After recovery from the spin, it can take a significant amount of time for the gyroscopes to re-orient themselves and once again provide accurate information. A tumble can be hard on the gyroscopic unit and reduce the life of that instrument. Some gyro instruments may allow you to cage the gyros prior to entering a spin, thus helping to reduce wear and tear on the instrument while doing spin training.

Even the ball in the turn and bank coordinator cannot be trusted to tell you the direction that the plane is spinning. Like gyroscopes, it is affected by the rolling, yawing, and pitching movements of the plane. If the ball is located on the left side of the cockpit, it will move to the left side of the tube in spins both to the right and left. Contrary to the old adage, "step on the ball," do not use the ball as an indicator of which rudder to push on during recovery. It could very well be wrong!

When I teach pilots spin recovery techniques, I stress how important it is to look OUTSIDE the plane. By doing so, you can learn which direction you are spinning and initiate the correct spin recovery inputs. Many pilots are at a loss when they first look at the instrument panel of the Pitts S-2B in which I instruct. There are no gyroscopic instruments in the plane; and for those who have become accustomed to staring intently at the artificial horizon and directional gyro, the bare panel in front of them is a new experience.

Keeping your head buried in the cockpit, and inputting the correct rudder only because you know that you started the spin in one direction and need to push in the other, is not the correct manner to recover from a spin. During lessons, I have students put their head down, and then I put the plane into a spin and have them recover. This reinforces the need to look outside the plane as they bring their head up to find out which way the plane is spinning.

Airspeed Indicator During a Spin

When your plane enters a spin, even though it is a maneuver with a relatively steep nose down attitude, the plane does not behave as it does in a normal dive. When in a spiral, the plane is diving and increasing in airspeed. In a spin, the plane may have a nose down attitude, but the airspeed indicator is not increasing. For this reason, the airspeed indicator will hover near the stall speed of the plane. While the speed may register near stall speed, there may be inaccuracies in the airspeed indicator's values, as there always are as the plane slows. This can also be coupled with an angle of attack that is at or above the critical angle of attack, causing even larger inaccuracies. If a plane goes flatter in its spin angle, the airspeed indicator may even drop toward zero as the angle of attack increases.

In my experience, flat spins have a slower airspeed and rate of descent than do normal spins. Pilots in flat spin training comment on the fact that they expected the maneuver to be more violent. But the flat attitude seems to keep the airspeed slower than during normal spins, making the maneuver seem less vigorous, at least from the aspect of rate of descent and airspeed. There are the occasional stories of pilots who have inadvertently entered a flat spin and are unable to recover from it. Due to the relatively low rate of descent found in flat spins, however, the impact with the ground is soft enough to allow the plane's occupants to walk away from the wreckage. Do not count on this being the case, though. While I have heard these tales on several different occasions, I have never actually talked to a pilot who experienced it firsthand.

ENGINE STOPPAGE

When in a spin, you should have the engine at idle, both for entering the stall and spin. As a result, the engine is turning low RPMs. For engines that turn in a clockwise direction of rotation, when you are spinning to the right, you are actually lowering the RPMs even further as the plane turns in the same direction as the engine's rotation. If the RPMs slow too much, the engine may not be able to continue to run. The greater the number of turns to the right during the spin, the greater the chances for the engine to quit. If it does stop, you need to dive the airplane to attain sufficient airspeed to get the propeller turning fast enough to restart the engine. If the plane you fly happens to have an engine that turns opposite the normal direction, doing spins to the left may cause the engine to quit.

EMERGENCY SPIN RECOVERY

Besides normal FAA-approved spin recovery procedures, there is another recovery procedure that was first publicized by Mr. Eric Muller and is now promoted by Mr. Gene Beggs. This procedure is based on findings that planes are able to recover from a spin if a simplified recovery procedure is used. This procedure reduces the pilot's need to correctly position the elevator and ailerons, or to know what type of a spin they are in. In short, the procedure is:

1. Cut that throttle!
2. Take your hands off the stick!
3. Kick full rudder opposite until the spin stops!
4. Neutralize rudder and pull out of the dive!

(*Sport Aerobatics*, p. 31, April 1994)

As you can see, this procedure differs from that recommended by the FAA. One major difference is in releasing the control stick, allowing the airflow over the controls to put the ailerons and elevator in the position they need to be for spin recovery. By inputting rudder opposite the spin, the plane recovers into a dive.

I have used this procedure successfully countless times to recover from normal spins, inverted spins, accelerated spins, and both upright and inverted flat spins in the Pitts S-2B I fly. I also teach this method of spin recovery for emergency spin situations. One of the advantages of this spin recovery method is that the pilot does not have to know the type of a spin to recover from it. In case you cannot determine the direction in which you are spinning, the correct rudder to push is the one that takes the greatest amount of force.

Mr. Beggs has tested the recovery procedure in several aircraft types with success, as long as they were loaded within proper weight and balance ranges. The aircraft tested by Mr. Beggs include several Pitts Special models, the Christen Eagle II, the Cessna 150, Cessna 172, and the Beechcraft Skipper trainer (*Sport Aerobatics*, p. 31, April 1994).

Certain cautions related to this recovery technique need to be mentioned though. NASA studies have found that this procedure does not work with all aircraft and should not be depended on in all cases. Some authors have suggested that the Muller/Beggs recovery procedure may work at certain points in a spin for some aircraft, while not working at other stages. Other NASA studies indicate that spin recovery cannot be achieved with rudder alone and require the correct input of rudder and elevator (*Sport Aerobatics*, p. 31, August 1994).

Spin rotation rates, angle of attack, or other factors may influence how well this spin recovery procedure works in a particular plane. If you are not certain it will work in the plane you are flying, contact the manufacturer. As always, the manufacturer is the final word in the correct spin recovery procedures for a given plane.

SUMMARY

The chapter began by defining spins and describing how aircraft are spin certified, as well as what bearing that has on whether a plane should be spun. Remember that just because a plane can spin does not mean it should spin. Normal category planes are prohibited from spinning. When you practice spins, use only planes that are certified for them, and that are loaded within the proper weight and balance ranges. If you fly a plane outside its weight and balance limits, even if it is certificated for spins, it may not be able to recover once it enters a spin.

The various stages of spins were also covered, with a review of the spin entry phase, the incipient phase, the developed spin phase, and the recovery phase. As you recall, it is easiest to recover from a spin while the plane is in the incipient phase, before the plane has settled down into a developed spin. The procedures outlined by the aircraft's manufacturer should always be used for spin recovery, but the FAA also has a recommended set of procedures.

Improper recovery procedures were also covered, with a discussion on how each control may affect spin development, and recovery. Lack of knowledge on the part of pilots entering spins may cause them to use controls incorrectly or in the wrong sequence. The sequence of control inputs is as crucial as making the correct inputs themselves. Follow the recommended spin recovery procedures for the plane you are flying in the correct sequence, and you will have much greater success during spin recovery.

Spins cause you to lose a great deal of altitude, and, if accidentally entered while in the pattern, offer much less chance of recovery before running out of altitude. The best spin recovery technique at low altitude is to never enter one. Stall/spin awareness and avoidance on your part is the best insurance you have in preventing low-altitude stall/spin accidents.

Instrument and engine behavior during a spin were also discussed. The main point is that a plane's instruments will behave in an inaccurate manner in a spin. Knowing that these abnormalities exist can help you feel less concerned when you experience them.

Finally, emergency spin recovery procedures as taught by Eric Muller and Gene Beggs were discussed. While I have had great success in using and teaching them, these procedures may not be successful for all aircraft. On the positive side, the use of these procedures greatly simplifies recovery procedures. Many accidental spin situations, especially during aerobatic flying, are the result of the plane entering a spin that the pilot cannot get out of because the type of spin is not recognized.

4

Normal Spins

CHAPTERS 4 THROUGH 8 PROVIDE AN IN-DEPTH EXPLORATION OF several different spin situations. This chapter includes a number of different views from aircraft as they enter a normal, upright spin. You will be able to visualize spin entry, the developed portion of the spin, and recovery from the spin. The views are from both the left seat looking forward over the nose and out the left window of the plane from the pilot's seat. This chapter is relatively short, given that much of the material related to normal, upright spins is covered in Chapter 3. Brief reviews of concepts are provided, but much of the chapter centers on the photographs and illustrations that give you a "pilot's eye view" of what is taking place during a spin.

After reading the chapter, you should be able to visualize and understand what is taking place during an upright spin. This knowledge should include not only the control inputs and how the plane will react to them, but the scenes you can expect to see from the plane during the spin.

SPIN ENTRY

To briefly reiterate concepts covered in Chapter 3, spins should always be done from a safe altitude that will allow recovery from the spin. When I am giving spin training, we normally begin spins no lower than 5,000 feet AGL. If multiple-turn spins are being executed, we climb even higher. While this may seem like an excessive amount of altitude, any wallowing during spin entry or delay in spin recovery can add a great deal to the amount of altitude lost during the maneuver. You should perform clearing turns prior to beginning the spin maneuver, to ensure that no other aircraft are in your vicinity. Always use carburetor heat as recommended by the aircraft's manufacturer prior to reducing engine power. Once you have cleared the area, reduce the power to idle and slow the aircraft to a stall while maintaining your heading. Maintain altitude as the plane slows by easing back on the elevator control until the plane stalls. Figure 4-1a shows the plane's attitude over the nose of the plane just prior to the start of the spin. Figure 4-1b displays the same view out the left window of the plane, looking down the wing. As you can see, the angle of attack relative to the horizon is relatively large as the plane approaches the stall. Note that all of the photographs in this chapter were taken in a Cessna 152. The angle may vary based on the plane you are flying, but the critical angle of attack normally falls in the 18 to 20° range (*Pilot's Handbook of Aeronautical Knowledge*, p. 9).

Rudder/Elevator/Aileron Positions

When the plane stalls, the ailerons should be in the neutral position to avoid putting the plane in an unusual spin attitude situation. Input full rudder in the direction you want to spin the plane. At the same time, input full nose up elevator, to help keep the plane in a stall as it enters the spin. These inputs should be smooth, brisk, and with authority, to help the plane quickly transition into the spin.

Figure 4-2a shows the view from the front of the plane as it begins to roll and yaw into the spin, which in this example is to the left. This picture clearly shows the steep angle of bank that the plane is in, with the nose of the plane beginning to drop well below the horizon. Figure 4-2b shows the view down

Fig. 4-1a Normal spin entry, just before stall, forward view.

Fig. 4-1b Normal spin entry, just before stall, side view.

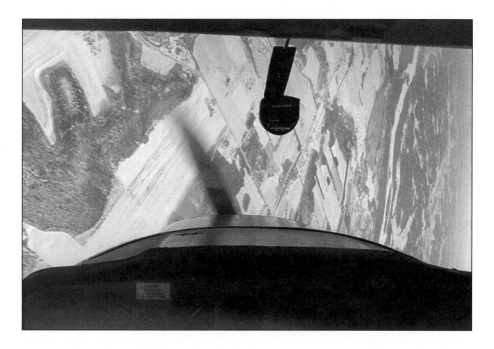

Fig. 4-2a Normal spin, just after spin entry, forward view.

Fig. 4-2b Normal spin, just after spin entry, side view.

the left wing of the plane, illustrating the angle of attack during this portion of the maneuver. As you can see, the left wing is well down at this point during spin entry.

The following series of photographs document the remainder of the spin as it develops from the initial stall and spin entry into a true spin. Figure 4-3a depicts how the nose has dropped to a vertical, descending attitude, actually passing through vertical, with the plane slightly inverted. Figure 4-3b, looking down the left wing at the same point, more effectively demonstrates this slightly inverted attitude at approximately one-half turn into the spin. The Cessna 152 regularly passes through this attitude during spin entry. The first time students execute spins in the Cessna 152, they are surprised by the extreme nose down attitude during not only spin entry, but also during the developed portion of the spin.

Figures 4-4a and 4-4b illustrate how the nose down attitude is reduced during the second turn of the spin, though the nose is still well below the horizon. The view forward over the nose of the plane still shows a considerable amount of real estate staring back at the pilot. The side view gives an excellent point of reference for the angle of the plane relative to the horizon. At this point, the plane is still in the incipient phase of the spin, but moving toward a developed spin. Figures 4-5a and 4-5b show the plane's attitude during turn three of the spin, which is now a developed spin. Again, the forward and side views show

Fig. 4-3a Normal spin, forward view.

Fig. 4-3b Normal spin, side view.

Fig. 4-4a Normal spin, second turn, forward view.

Fig. 4-4b Normal spin, second turn, side view.

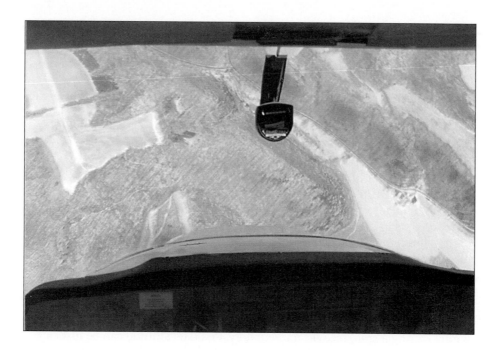

Fig. 4-5a Normal spin, third turn, forward view.

Fig. 4-5b Normal spin, third turn, side view.

how the plane is still in an extreme nose down attitude. The nose down attitude and rate of rotation may vary during the spin.

Disorientation During the Spin

Pilots just beginning spin training may not be sure of the plane's attitude during a spin. The extreme nose down angles, coupled with the plane's rapid rotation, may make it difficult to understand what the plane is doing. Often new spin students feel that they have gone through several turns when, in fact, the plane has only completed one turn. To help reduce this disorientation, I have pilots pick a point on the ground and use it as a reference while they are executing a spin. During the spin, they then count each time the point passes the nose of the plane. This technique helps them maintain a reference during the spin and aids in keeping track of the rate of rotation as the spin progresses.

Another important consideration when doing spins is keeping track of the plane's altitude. Pilots tend to focus on one or another aspect of the spin, not noticing other aspects of what is taking place. Attention to altitude is normally one of these areas of neglect. During a spin, I remind students to frequently glance at the altimeter. Before beginning a spin, decide what altitude you need to recover at, and then recover before you reach it. If things ever start to go wrong during a spin, recover immediately. Precious altitude can be lost as you focus on the spin. I know pilots who have come dangerously close to not having sufficient altitude to recover from a spin that has gotten out of hand. If they had noticed that the spin was getting out of hand and recovered immediately, these close calls could have been avoided.

Common Spin Entry Errors

Several improper spin entry scenarios are covered in Chapter 3, but these situations are summarized again in this chapter. First, maintain ailerons in the neutral position. Recall that pro-spin aileron input can cause the plane to enter a "knife-edge" spin, while anti-spin ailerons can cause the spin to flatten. Pilots often use pro-spin ailerons as they enter the spin. They may enter pro-spin aileron in an attempt to maintain coordination with the rudder's input.

In addition, full rudder input should be used when entering a spin. Pilots occasionally use only partial rudder input, which causes the plane to wallow as it enters the spin. In some cases the plane may not enter a spin at all, because the wings are not sufficiently stalled. Full up elevator should be added as the plane stalls for the same reasons. Having less than full up elevator can cause the plane to recover from the stall and not enter the spin as a result of the change in the plane's angle of attack as the plane pitches down after the stall. In certain cases, it may wallow before entering the spin; while in others, the plane will enter a spiral or a dive. When the plane wallows before finally entering a spin, excessive altitude can be lost.

Power should be at idle as you stall the plane. Use of any power during the stall or spin may cause the spin to flatten, making it difficult, if not impossible, to recover. The first step in the FAA's recovery procedure is to reduce power to idle, and many aircraft manufacturers also have this as the first step in spin recovery.

SPIN RECOVERY

Spin recovery from normal, upright spins was also discussed at some length in Chapter 3. You should always use the spin recovery procedures recommended by the aircraft manufacturer. The FAA-recommended spin recovery procedures will work for many single-engine, general aviation aircraft:

1. Retard power.
2. Apply opposite rudder to slow rotation.
3. Apply positive forward elevator movement to break stall.
4. Neutralize rudder as spinning stops.
5. Return to level flight.

(*Flight Training Handbook*, p. 157).

Remember that these steps must be completed in the correct order. Executing the recovery steps in the wrong sequence can result in an unsuccessful spin recovery as well as worsening the situation. A normal, upright spin can turn into an accelerated spin or an inverted spin when spin recovery controls are applied in the wrong sequence. Both of these situations are covered in Chapters 5 and 6. The FAA also cautions that brisk control inputs should be used during spin recovery. Use of slow control inputs may not result in spin recovery, even when the controls are at the stops.

Common Errors

This section briefly recaps the common errors discussed in Chapter 3. Remember not to retard the throttle after entering a spin, which increases the possibility that the plane may enter a flat mode due to the nose up moment that may result. Also, do not use full rudder opposite the direction of the spin to stop the spin's rotation. Pilots occasionally input only partial rudder opposite the spin during the recovery. When less than full rudder opposite the direction of spin rotation is used, it may take longer for the spin's rotation to stop. In some cases, the plane may not stop rotating until vigorous, full rudder is applied opposite the direction of the spin's rotation.

Once the rotation has stopped, you must then remember to neutralize the rudder. If you continue to hold counter-spin rudder, the plane may begin to yaw, and potentially spin, in the other direction. Even if the plane does not re-enter a spin, the plane will be yawing badly during the remainder of the recovery. If you find that you are being thrown to the side of the plane after recovery, it is quite possible you have not remembered to center the rudder. This may sound like a simple thing to remember, but spin students occasionally fail to center the rudder during their first few spins.

Elevators also play a crucial role during spin recovery. Once the spin's rotation has slowed, forward elevator control should be applied, which decreases the wings' angle of attack to less than the critical angle of attack. Depending on the plane, a brisk forward movement of the elevator control may be necessary. As they recover from a spin, some pilots forget to release elevator backpressure, thus causing the plane to remain in the stall. Other pilots jam the elevator control full forward, causing the plane's occupants to hang in their seat belts and the plane to dive at a steep angle. Both of these actions cause the plane to lose excessive altitude before spin recovery is completed.

Ailerons should be maintained in the neutral position during the spin. Use of counter-spin ailerons can cause the plane to enter a flat mode, and use of pro-spin ailerons may result in an increased rate of rotation, or a "knife edge" spin, with the plane steeply banked as it spins.

Unless otherwise recommended by the plane's manufacturer, flaps should be retracted when a spin is entered. Extended flaps can cause the plane to enter a flat mode and may make it more difficult to recover from the spin, or may cause structural damage to the aircraft.

Diving after the spin stops is another problem pilots may encounter, with the plane gaining excessive airspeed and losing more altitude than necessary during the recovery. The excessive airspeed can place undue stress on the plane and result in your pulling more g's than the plane's design allows. This can also result in an accelerated stall and potential damage to the plane. You should always avoid secondary stall/spins, as well as slow spin recovery. As with diving after spin recovery, both of these situations can result in your losing more altitude than you should.

AIRCRAFT CHARACTERISTICS DURING A SPIN

Each plane exhibits different flight characteristics while in a spin. One model aircraft may spin very well, easily allowing students to enter and exit spins. Other aircraft may easily enter spins but stubbornly refuse to recover from them. Then there are planes that resist the pilot's attempts to cause them to enter a spin. An engineer involved with one single-engine, general aviation trainer noted in an NTSB accident report that the prototype demonstrated very desirable spin characteristics, but changes made to the fuselage in the production model resulted in a plane that was much more difficult to recover from a spin.

Changes in center of gravity, pitch attitude during spin entry, and many other factors can affect how the plane will act during a spin. On one flight in a Cessna 152, I attempted to make the plane enter a spin from a slipping approach stall with a student in the left seat. In a power off configuration, it only wallowed in a stall, never entering a spin. When I executed the same maneuver with only myself in the plane, it readily entered a spin from that same flight mode.

Once in a spin, many aircraft do not remain steady state, but vary their spin characteristics. The next section examines the potential changes in the nose down pitch angles and rate of rotation that a plane may exhibit during a spin. Once you have completed this section, you will be better prepared to notice the changes that the plane you are spinning may display.

Pitch Oscillation

You may be surprised that once a plane is in a spin, it does not normally maintain a constant nose down attitude or rate of rotation. Many planes exhibit a rising/falling pattern in the pitch attitude as the plane spins. NASA reports have documented that test aircraft exhibit this change in pitch attitudes during the course of the spin. In my own experience, Pitts S-2Bs demonstrate a repeating pitch pattern during which the nose oscillates between a low and high point as the spin progresses. The nose of the plane reaches the lowest pitch down attitude at the 180° point after spin entry. As the spin's rotation continues, the nose then begins to rise again, to a high point at 360° of turn.

In a normal spin, the nose is always well below the horizon and this nose high/low pattern continues through the entire spin in this plane. When you practice spins, notice how the airplane you are flying behaves and what pitch attitudes it passes through during the spin. Being aware of the plane's pitch attitude during spins can help you avoid trouble situations. NASA reports have documented that planes are more difficult to recover from spins where the nose is higher, or in a flat mode. You should always recover from a spin immediately if the plane starts to enter a flat mode, where the nose rises higher than is normal for a spin in that plane.

Rate of Rotation Variation

I have found that the rate of rotation slows in the Pitts S-2B as the nose rises, then increases as the nose falls, through the course of a spin. Students receiving flat spin training have commented on the fact that the plane seemed to spin more slowly while in a flat spin than in a normal spin. While every airplane behaves differently in a spin, if you notice the rate of rotation is slowing, then the plane may be flattening out in the spin. An aft center of gravity or extended flaps can cause the plane to enter a flatter mode during the spin. Even if the plane is within its center of gravity and the flaps are retracted, some aircraft are more susceptible to flat spins than others.

One NTSB accident report documents that a particular aircraft had the nose rise well above the horizon during spin testing and one of the occupants was certain that it would never drop below the horizon. To help avoid getting into an accidental flat spin, listen to the plane and watch the horizon as you execute spins. You will notice a change in the sound of the air flowing over the plane as the rate of rotation and pitch changes. If you begin to see the nose rise, the rate of rotation slow, or hear changes in the sound of the plane, you may well be entering a flat spin. In this case, be sure to recover immediately.

SUMMARY

The factors affecting an upright spin were discussed. The chapter covers how to enter and recover from a spin, and includes photographs that illustrate what you can expect to see during a spin. These pictures should help prepare you for the attitudes and views you may find when you practice spins.

I recently flew with a new student. During our preflight discussion, this pilot said that the first time she was exposed to a spin with another instructor, the nose down attitude was very startling. The nose down attitude, combined with the rapid rate of rotation, quickly caused this individual to lose track of what the plane was doing.

When giving spin training to students, I call off every half turn during the spin. This helps students keep track of where they are during the spin. Review the pictures in this chapter in preparation for spin training. The forward and side views should give you an idea of what you can expect as you spin a plane. Keep in mind that you may need to look around to find the best place to view the horizon. To determine the direction of rotation, the best place to look in most cases is directly over the nose of the plane.

Spin entry and recovery procedure were summarized. Common errors associated with spins were also reviewed. Always use the spin recovery procedures recommended by the aircraft's manufacturer. In one NTSB spin accident report, a pilot reported that the aircraft in question had been difficult to recover from a spin. When spinning the plane involved in the accident, that pilot found that the FAA-recommended spin recovery techniques would not recover the plane from a spin. Having tried these procedures while in a spin, the pilot found that only the spin recovery procedures documented in the plane's operations manual would successfully recover the plane. For this reason it is very important to know the proper spin recovery procedures for the plane you are flying.

The next few chapters look at more exotic spins, which are spins that you may find yourself in if you do not correctly recover from stalls or upright spins. Many stories and myths surround these spins, such as the flat spin. You may be completely unaware that some of these spins even exist, but knowledge of what causes them can help you avoid them.

5

Accelerated and Crossover Spins

THE NEXT FEW CHAPTERS DISCUSS SOME OF THE MORE UNCOMMON spins a pilot can encounter. Pilots have built many myths around some of these spin sequences, though others may be completely foreign to you.

One of the problems that aerobatic pilots occasionally run into is unplanned spin entry while executing an aerobatic maneuver. Over the years I have read accounts of pilots in airshows spinning into the ground while performing what had been seemingly normal aerobatic routines. As an aerobatic flight instructor, I have had students place me in countless different spin situations while attempting to execute everything from a loop to a hammerhead. These unplanned spins often do not fall into the normal, upright spin family. Safety dictates that aerobatic pilots must understand what some of these more exotic spins are, be able to recognize them, and know how to recover from them.

Even if you are not flying aerobatics, knowing how these spins can be initiated can help you avoid ever entering them. Many aircraft that are certified for normal, upright spins will not be able to perform the spins described here safely. For this reason, to get training in these maneuvers you should find a flight school with the proper aircraft and flight instructors. The more I learn about spins, the more convinced I am that pilots who properly practice spins are safer because of their experience. Under normal flying conditions, spin avoidance is the best defense you can have to stay out of dangerous situations. This chapter describes accelerated and crossover spins and how they are entered. This knowledge alone may help you avoid these spin situations.

The accelerated spin can develop out of any spin scenario, whether it is a normal upright spin, an inverted spin, or flat spin, either upright or inverted. Once entered, an accelerated spin can become disorienting as the rate of rotation increases. Along with this higher rate of rotation comes an increase in the associated g-forces, positive or negative, depending on the spin orientation.

Crossover spins can develop from accelerated spins and can be even more disorienting than an accelerated spin. The first step to entering a crossover spin is ending up in an accelerated spin, then taking the wrong spin recovery steps. The figures accompanying the explanations will help to give you a feel for attitudes the plane passes through as it is executing each spin.

ACCELERATED SPINS

Until I became involved in aerobatics, I had never heard of an accelerated spin, even though I had done many spins since I began to fly. But the first time I experienced one several years ago is as clear in my mind as the day it happened. In short, an accelerated spin can begin as a normal spin. For most pilots, this means an upright spin, though they can also develop out of inverted spins and flat spins. This explanation assumes the pilot starts out in a normal, upright spin.

Figure 5-1a shows the control positions of a plane that is in a normal, upright spin to the left. The control stick is in the full aft position, full left rudder has been applied, and the throttle is at idle. At this point our unwary pilot breaks one of the cardinal rules of spin recovery, by not making the recovery

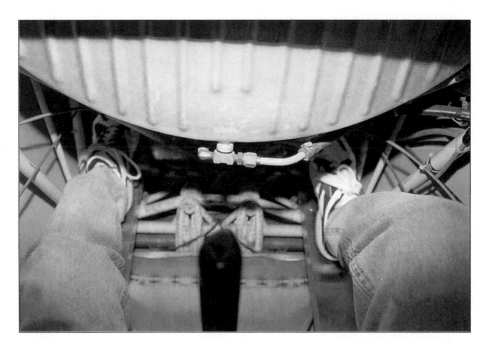

Fig. 5-1a Control positions left upright spin: left rudder, aft stick.

inputs in the correct order. Remember that the FAA spin recovery procedures require this sequence:

1. Retard power.
2. Apply opposite rudder to slow rotation.
3. Apply positive forward elevator movement to break stall.
4. Neutralize rudder as spinning stops.
5. Return to level flight.

(*Flight Training Handbook*, p. 157).

We will assume that these recovery procedures work for the plane in the example and that the pilot has pulled the throttle back to idle. Now, instead of using full opposite rudder to start recovery, he instead applies forward elevator input, expecting to break the stall. Figure 5-1b shows the new control configuration: engine power is at idle, full left rudder is still being maintained, but the control stick is pushed forward. The pilot thinks that once the stall is broken, right rudder can be used to stop the yaw. Unfortunately, this is not what happens. As forward elevator pressure is applied, the plane's angle of attack changes, just as you would expect. However, the result of this angle of attack change is a dramatic increase in the spin's rate of rotation, not a recovery from the stalled condition.

While not a completely accurate analogy, picture an ice skater during a pirouette. She begins this "spin" with her arms extended, and then draws her arms in closer to her body. As the skater brings in her arms she begins to spin

Fig. 5-1b Control positions accelerated spin: left rudder, forward stick.

faster. Assume the skater's action of drawing in her arms is the equivalent of the forward input on the elevator control, changing the plane's angle of attack. Like the skater, the plane begins to accelerate its rate of rotation as the pilot pushes further forward on the control stick. What started as a docile, normal spin suddenly turns into something unexpected.

I have been unsuccessful in finding a study or satisfactory explanation of what causes this increased rate of rotation. My theory is that as the forward elevator input forces the nose of the plane down, the change in the angle of attack increases the difference in the lift being generated by the wings. Recall that the wing to the outside of the spin is generating more lift than the wing to the inside of the spin. As a result of this greater lift differential, the rate of rotation increases. Figure 5-2 illustrates this increase in lift differential.

The first time I experienced an accelerated spin was with an aerobatic instructor in a Pitts Special. I had done many spins over the years and was

RAISED WING HAS NOT STALLED AND IS PRODUCING GREATER LIFT THAN STALLED WING

LOWER WING HAS PASSED CRITICAL ANGLE OF ATTACK AND IS PRODUCING LITTLE LIFT

NORMAL SPIN LIFT DIFFERENTIAL

RAISED WING IS NOW GENERATING EVEN GREATER LIFT THAN STALLED WING

ACCELERATED SPIN LIFT DIFFERENTIAL

Fig. 5-2 Lift differential in accelerated spin.

unconcerned with what this spin would be like. We climbed to a safe altitude and did all the proper clearing turns. I started an upright spin, then pushed the control stick forward as I was instructed to do. Wow! I was amazed at how quickly the rate of rotation increased. The Pitts Special is a great little airplane to spin, and knowing just how capable it is was a confidence builder. But I was completely unprepared for how fast that little airplane started to turn. I executed the proper recovery procedures for that plane (no longer the FAA-recommended procedures at that point), and the plane responded accordingly.

After that flight I gave a great deal of thought to how unsafe this spin could be if it had taken place in a plane other than a highly capable aerobatic airplane. Furthermore, if the pilot did not know how to recover, the spin could clearly become dangerous quickly.

The photographs of accelerated spins in this chapter use the familiar forward and side views. Look at the pictures taken at the beginning of the normal spin and contrast them to the angle of attack in the accelerated portion of the spin. The pictures were taken from a Pitts Special S2-B. Because you could overstress a normal airplane executing this spin, do not attempt it unless your plane is capable of executing the maneuver safely.

In the pictures that are views to the left, you will notice an object resembling a diamond trailing from the I-strut between the upper and lower wings. This is a sighting device that helps pilots align both vertical lines and 45° lines when flying aerobatics. For the purposes of discussions in this book, the device makes the attitude of the aircraft easier to understand. In some pictures readily discerning the attitude of the plane may be difficult at first. This is due to the very unusual attitudes the plane is in during the different spins. Imagine how much more difficult it would be to be in the plane, attempting to fly the maneuver, and understand the attitude of the plane. The need to know just where to look to find the best view becomes very important in these situations.

Spin Entry

Spin entry into an accelerated spin begins with the same entry you know from normal, upright spins. You should compare the views of normal spin entry from the Pitts Special, which are included in this chapter for comparison to accelerated spin views from the same plane, to the Cessna 152 normal spin entry photographs in Chapter 3. Figures 5-3a and 5-3b show the forward and side views to the left as the plane stalls. Figures 5-4a and 5-4b show the plane just as it begins to roll and yaw into the spin. Note the angle of bank and the wing's angle of attack at this point during spin entry.

As you can see from the sequence of pictures, the plane was stalled, and then put into a spin to the left. Full left rudder, full aft elevator, and neutral ailerons were used as the plane stalled. The throttle was at idle. Incorrect use of controls during spin entry will result in the same spin problems noted in Chapter 3. At this point in the spin, the controls are as depicted in Figure 5-1a.

Fig. 5-3a Normal spin to left at stall: forward view.

Fig. 5-3b Normal spin to left a stall: left side view.

Fig. 5-4a Roll into spin: forward view.

Fig. 5-4b Roll into spin: left side view.

Accelerated Spin Characteristics

Figure 5-5a shows the view over the nose of the plane at approximately two turns into the spin. The plane has now settled into a developed spin. Figure 5-5b shows the view to the left at the same point of the spin. To begin the accelerated spin, full left rudder and neutral ailerons are maintained as the control stick is pushed briskly forward. Figures 5-6a and 5-6b show the forward and side views from the plane just as the control stick reaches the full forward position. Note the change in the plane's pitch angle from the previous set of figures when the plane was in a normal spin.

The plane begins to increase its rate of rotation as the elevator control is moved forward. With this increase in spin rotation comes an increase in positive g-forces, as well. The rate of rotation has increased a great deal by the time the control stick is full forward. Prior to teaching accelerated spins to pilots, I give them an extensive explanation of what to expect as they input forward control stick. Even with this prior knowledge of what is to come, every student I have given accelerated spin training to is not prepared for how the plane reacts to this seemingly innocent forward movement of the control stick. And in the fictitious example earlier in the chapter, the problems began because the pilot did not think the order of spin recovery procedures was important. As you

Fig. 5-5a Two turns into spin: forward view.

Fig. 5-5b Two turns into spin: left side view.

Fig. 5-6a Control stick full forward: forward view.

Fig.5-6b Control stick full forward: left side view.

will recall, the pilot was attempting to break the stall with forward elevator before the rotation was stopped with full, counter-spin rudder.

Moving on to inverted, accelerated spins, entry begins as an inverted spin. These spins are covered in Chapter 6; but for now, assume you have already established yourself in an inverted spin. In this type of spin, the control stick is held full forward to maintain the stalled condition. So, you are inverted, the throttle is at idle, full left rudder was applied as the plane stalled, and you are holding the stick full forward. Under normal circumstances, this would put you in an inverted spin to your left with about two negative g's pushing you away from the seat. Pilots must often lean forward, with the arm holding the elevator control fully extended, to keep the control stick all the way forward during the spin.

When pilots first experience inverted spins, the physical forces can quickly disorient them. A typical response to this very unusual attitude is for the pilot to relax back into the seat and allow the stick to move back from the forward stop. Because you are inverted, this is the equivalent of pushing the control stick forward in an upright spin. The plane's angle of attack changes and the rate of rotation increases. Inverted spins have a higher rate of rotation than upright spins to begin with; and as the stick moves back toward the pilot, the rate of rotation increases further. This, of course, increases the negative g's and adds to the disorientation the student is feeling.

Accelerated spins are also possible while in both upright and inverted flat spins. In each of these spin modes, the rate of rotation increases as the position of the control stick changes. In an upright flat spin the rate of rotation increases as you move the stick forward, just as with the normal, upright spin. This increased rate of rotation can raise the nose even higher in relation to the horizon than in an unaccelerated upright flat spin. If the control stick is moved back toward the pilot while in an inverted flat spin, the spin's rate of rotation will increase then as well.

At this point it should be apparent that accelerated spins can take place from any spin situation. The importance of executing spin recovery procedures in the correct order should also be very clear. For this reason, every pilot—not just aerobatic pilots—needs to be aware of accelerated spins and what causes them. You can then take steps to avoid entering one of these spins.

Recovery from Accelerated Spins

As a caveat to what follows, I must mention that I have used the accelerated spin recovery procedures only in Pitts Specials. There is no guarantee that the procedures will work for any other plane. Before you attempt an accelerated spin, be certain you know that the plane you are flying is capable of executing and recovering from an accelerated spin. In addition, because the plane you fly may require a different set of recovery procedures, you should know those procedures as well.

I have used the Muller/Beggs emergency spin recovery steps as the standard procedure for recovering from accelerated spins of all types. This sequence was covered in Chapter 3; but to review, the correct steps are:

1. Cut that throttle!
2. Take your hands off the stick!
3. Kick full rudder opposite until the spin stops!
4. Neutralize rudder and pull out of the dive!

During spin recovery practice, I have students put the plane into an accelerated spin and then recover from it. To reinforce the correct recovery procedures, I have each student say the procedure as it is executed. In addition, I also have them hold their right hands above their heads as they release the control stick to reinforce the need to let go of it.

I attended an aerobatic safety seminar at which one speaker, an airline pilot who also competes in aerobatics, explained that he had not believed the Muller/Beggs technique really worked for spin recovery. Like many pilots, he had felt the plane could not do a better job of spin recovery with his hands off the controls than if he actively controlled the spin recovery.

Then he told the tale of how one day he was executing an inverted flat spin. As the spin progressed, he relaxed the arm that held the control stick forward, allowing the stick to move back toward him. As noted earlier, this will cause an inverted flat spin to turn into an accelerated spin. The speaker tried repeatedly to recover manually from this accelerated, inverted flat spin using hands-on recovery methods, but the plane did not respond. Finally, running out of altitude, the pilot decided it was time to get out of the plane. He let go of the control stick to unbuckle his safety harnesses. Before he completed unfastening the harness, he looked up and discovered the plane had recovered from the spin on its own. During the process of working with the harness, he had been holding counter-spin rudder. When he released the control stick to release the seat belt, the plane recovered. This was a hard-learned lesson, but this individual now teaches the Muller/Beggs recovery method to his own students.

During a recent lesson, one of my students put the plane into an accelerated inverted spin. We hadn't planned to perform this maneuver, but as fate would have it, we ended up in one anyway. While in the spin, there was a significant resistance to movement of the control stick for elevator control until counter-spin rudder was applied. Once counter-spin rudder was applied, the stick moved of its own accord to the proper position, and the plane quickly recovered from the spin.

This was a valuable learning experience for both me and the student. Over the years I have executed quite a few accelerated spins, both planned and unplanned. I have always accepted the Muller/Beggs spin recovery procedure as the proper method for accelerated spin recovery. I had never tried to move the control stick while in an accelerated spin, and I took the opportunity to experiment during this particular spin. After having tried to center the stick, it was easy to see how getting the control stick to just the right position for recovery in this situation could be difficult. The stick forces and g-loadings, combined with the motions of the plane, may make it difficult to identify exactly the right position for the control stick. The previously mentioned speaker at the aerobatic safety seminar felt that in these situations, positioning the control stick only a little off the correct position could prevent spin recovery.

Figure 5-7a shows the forward view from the plane after the control stick has been released and counter-spin rudder applied. Figure 5-7b shows this same point in spin recovery looking to the left. At this point the plane is in a normal spin recovery attitude and can be flown back to straight and level flight.

Common Errors

All of the common spin recovery errors that were discussed for normal spins apply to accelerated spins. The most frequent error that applies to the Muller/Beggs spin recovery technique is to fail to release the control stick. Many pilots express serious misgivings about the plane recovering on its own and have a tendency to hold onto the stick during the recovery. In some cases they are merely following along as the stick moves of its own accord, which does not hinder recovery. But those who are actively attempting to center the control stick may slow or prevent the plane from recovering. A common error I have witnessed related to spin entry is a hesitancy for pilots to push forward on the control stick to begin the accelerated spin. Having some idea of what is going to take place when they push the stick forward, they are not completely comfortable with the idea of purposely starting an accelerated spin. In the proper aircraft and with a qualified instructor, accelerated spins are as safe as any other spin.

Fig. 5-7a Accelerated spin recovery: forward view.

Fig. 5-7b Accelerated spin recovery: left side view.

CROSSOVER SPINS

Accelerated spins are an interesting maneuver to practice in a planned, controlled situation. As noted in the last section, executing spin recovery procedures in the wrong sequence can have what could be called a negative impact on your ability to recover from a spin. Pushing forward on the control stick while holding pro-spin rudder puts the plane in an accelerated spin. This section takes that incorrect sequence of spin recovery procedures one step further. There are two reasons to do so. The first is to make you aware of yet another spin situation, the crossover spin. The second is to finish underscoring the need to execute spin recovery steps in the proper order. By continuing the improper sequence of recovery procedures while in an accelerated spin, you can cause the airplane to transition from an upright, accelerated spin to an inverted spin. When this happens as an unplanned situation, your spin scenario can go from bad to worse!

Defining Crossover Spins

Figure 5-8a shows an imaginary plane in an accelerated, upright spin turning to the left. The nose is pointed well down and the rate of rotation is very high. Figure 5-8b shows the controls at this point, with the control stick forward, the ailerons centered, and full left rudder input. Our unsuspecting pilot, now faced with an accelerated spin after pushing the control stick forward, is operating under the assumption he only needs to make the correct spin recovery control movements. Even after entering the accelerated spin, this pilot fails to understand that the order of recovery steps is important. At this point in the spin, the pilot is concerned, not quite sure why the plane is spinning even faster, so he takes another step in spin recovery, pushing full counter-spin rudder. As noted in the previous section, to recover from a spin to the left the pilot uses full right rudder. Figure 5-8c illustrates the controls, showing the use of full right rudder.

In the case of an accelerated spin, counter-spin rudder does not stop the rotation of the spin. Instead, this causes the plane to flip over on its back and enter an inverted spin. Once the plane has crossed over from the upright spin to the inverted spin, the forward stick and full right rudder become pro-spin inputs. These control inputs cause the plane to remain in this very new situation, an inverted spin. Our intrepid pilot has just executed a crossover spin. This transition can take place so quickly that in some cases pilots do not even realize what has taken place, or that they are now in an inverted spin. Figure 5-8d shows the plane in an inverted spin.

As always, not all planes behave the same when in a spin, and even the same plane can vary from flight to flight in how it reacts to these situations. The crossover spin is more demanding of the pilot and plane than any spin previously discussed. For this reason, be certain that your plane is capable of executing it safely, that you are within weight and balance restrictions, that you have lots of altitude, and the instructor teaching you is absolutely comfortable teaching them.

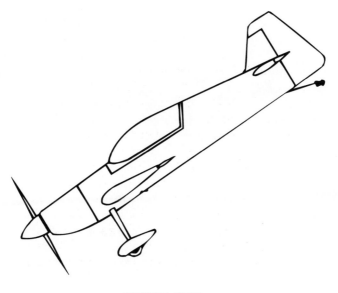

NORMAL SPIN

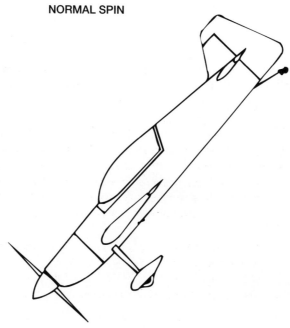

ACCELERATED SPIN

Fig. 5-8a Plane in accelerated spin: outside image.

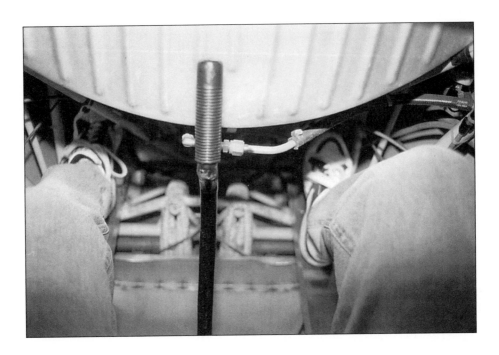

Fig. 5-8b Control inputs for accelerated spin.

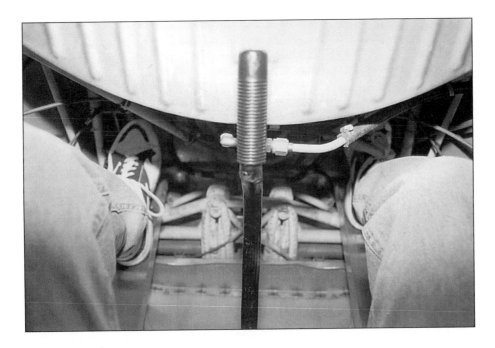

Fig. 5-8c Control inputs for accelerated spin: full right rudder.

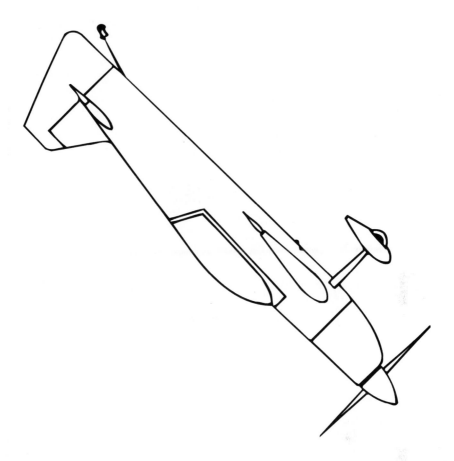

Fig. 5-8d Inverted spin: outside image.

Spin Entry

The spin entry is done using the same procedures as any normal, upright spin. The one difference is you should have as much altitude as you can get. I like to start crossover spins at 9,000 feet AGL or more. During this maneuver, the plane has the potential of losing a tremendous amount of altitude. The last thing you need to worry about as you recover is whether you have enough altitude to get the job done. This example uses a spin to the left so as the plane stalls, full left rudder is applied and the stick is held fully aft.

Once the plane is approximately two rotations into the spin, begin pushing the control stick forward, keeping the ailerons centered. You should also maintain full left rudder. This will cause the plane to enter an accelerated spin, with the plane's rate of rotation increasing. Figure 5-9a shows the view over the nose of the plane at this time, while Figure 5-9b shows the view to the left.

Fig. 5-9a Accelerated spin: forward view.

Fig. 5-9b Accelerated spin: left side view.

I then typically input full right rudder after approximately two turns into the accelerated spin. The right rudder input causes the plane to transition from an accelerated, upright spin to an inverted spin. The transition can be a wild ride, so be prepared for what might take place! Figures 5-10a and 5-10b show the forward and left side views from the plane once it has completed the crossover and is now in an inverted spin.

Improper Spin Entry

Spin entry problems for normal spins also apply in this situation. In addition, some pilots start pushing the control stick forward to begin the accelerated spin while the plane is early in the incipient phase. Occasionally, this causes the plane to recover rather than enter an accelerated spin. Another problem is not applying full opposite rudder after the accelerated spin has started. This can cause the plane to continue in the accelerated spin, not making the crossover to an inverted spin. You should also be sure to maintain neutral ailerons as you move the control stick forward and while you are making rudder inputs. If at any point during this spin, or any other spin, things begin to get out of control, execute spin recovery immediately.

Spin Recovery

As with accelerated spins, I have students use the Muller/Beggs procedures to recover from a crossover spin once the crossover is completed and the plane is

Fig. 5-10a Inverted spin after crossover: forward view.

Fig. 5-10b Inverted spin after crossover: left side view.

in an inverted spin. When you begin in an upright spin to the left, you will have initially input left rudder as you entered the spin. Left rudder will be maintained in the accelerated portion of the spin. To initiate the crossover spin, you enter full right rudder. Therefore, to recover from the inverted portion of the spin, you input full left rudder to stop the rotation. (This is still your left, even though you are now upside down.) Figure 5-11a shows the forward view after the inverted spin recovery. The view to the left, Figure 5-11b, shows the attitude that can clearly be seen at this same point. You recover in an inverted attitude and will need to roll to an upright attitude. Remember to follow the correct sequence for the Muller/Beggs spin recovery procedures.

Disorientation During the Spin

The spins covered so far in this book—the upright spin, the accelerated spin, and the crossover spin—have been presented in order of increasing complexity. With each step up in complexity, the potential level of pilot disorientation while in the spin has also increased. In the crossover spin you are going from an upright spin to an inverted spin in not much more than the blink of an eye. Pilots who have accidentally executed a crossover spin have told me how it took one or two turns after the crossover before they realized they were upside down. This may seem difficult to believe, but when you are in an accelerated spin and not able to recover from it, slipping over to inverted can escape your attention because you are focusing on trying to get the plane out of the spin.

Fig. 5-11a Inverted spin after recovery: forward view.

Fig. 5-11b Inverted spin after recovery: left side view.

During the accelerated portion of the spin, you will be pulling higher positive g's than in a normal upright spin. This, combined with the rapid rotation, can make it difficult to keep track of how many turns the plane has completed in the spin, as well as the plane's attitude. As you cross over from upright to inverted, you will then be pulling negative g's. Be sure to relax your body as the plane becomes inverted. Depending on how rapid the crossover is, you may pull between 2 and 4 negative g's initially as the airplane slips into inverted flight.

The first time I executed a crossover spin, I was tensing against the positive g's in the accelerated portion of the spin, but did not relax as the plane began to pull negative g's after the crossover. In this particular spin I pulled about 3.5 negative g's. The result of the continued tensing after the negative g's started was that I increased the intervascular pressure in my head and ruptured blood vessels in the whites of my eyes. There was no eye damage and no pain, but I was scary looking until they healed.

Be sure to have your lap belt as snug as possible when executing this maneuver. Nothing feels worse than hanging against the seat belt, with your derriere several inches off the seat cushion because the safety harness is not cinched tight enough. Depending on the plane you are flying and how much headroom it has, you may also find your head uncomfortably pressed against the canopy. This can add to any discomfort you may experience during the crossover spin and is a distraction you do not need as you are recovering from the spin.

Students often mention having difficulty determining the direction of rotation of a spin. You will find the best place to look to figure out which direction the plane is spinning is directly over the nose of the plane. Then, push rudder opposite the direction of rotation. If you look in different places, you may not be able to determine the plane's direction of rotation as easily. Many pilots look above their heads during inverted spins—in other words, up at the ground—to try and figure out which way they are turning. In many cases, this only serves to increase the disorientation they are experiencing. However, if you find that looking somewhere other than over the nose of the plane works best for you, then use that technique.

Common Errors

The most common error I have seen during crossover spins is that students do not push full counter-spin rudder to recover from the inverted spin after the crossover takes place. Many times with normal, upright spins, a plane will recover from the spin with only partial counter-spin rudder input. Do not become complacent about this! In these more advanced spins, and even in the upright spin, lack of full rudder input may not even slow the spin's rotation. I have had to remind a number of students to input full counter-spin rudder after they have input only partial counter-spin rudder.

Another common error involves not knowing which way the plane is spinning, and consequently which rudder to use for recovery. Students become disoriented as the plane transitions from a normal spin to an accelerated spin to an inverted spin. As a result of the changes taking place, they are unsure as to which rudder they should use for recovery. It does not matter if the plane is upright or inverted for recover rudder; use full, brisk rudder opposite the direction of the spin.

In some cases, students have not been able to determine which direction they were spinning, and for that reason are uncertain about which rudder to use. Looking over the nose of the plane should give you some indication as to the direction you are spinning. If the situation becomes critical, and you are not able to figure out which direction you are turning, try pushing against the rudder pedal with the most resistance against it. If this does not recover the plane from the spin after several turns, try the other rudder.

Some aircraft can take a number of turns to recover from a spin. Do not assume that if the recovery is not immediate, you are using the wrong rudder. In an emergency spin situation, remember to keep thinking and do not panic. Freezing or taking random actions to get out of the spin is the last thing that will help you recover from a crossover spin and goes against the maxim that is often stated by the FAA: "Keep flying the airplane."

Another common error is to fail to center the rudder after you have recovered from the inverted spin. After the excitement of the ride of a crossover spin, students can forget to neutralize the rudder once the inverted spin's rotation has stopped and the stall has been broken. As with any other spin, this can cause the plane to yaw badly after the spin has stopped, or enter an inverted spin in the opposite direction. In fact, students often brace their feet against the rudder pedals during inverted flight to help keep their legs from dropping toward the instrument panel. (When you are inverted, if you relax your legs they will drop down or toward the instrument panel.) This increases the chances you will fail to center the rudder, so be sure to give thought to what to do with your feet and the rudder pedals as you recover from the inverted spin.

Finally, remember to roll to upright after the spin stops. The roll is the quickest way to get to upright flight. Occasionally pilots operate under the mistaken assumption that by executing a half loop, they can easily get the plane back to straight and level flight. This will definitely lose more altitude than rolling to upright and may cause the plane to gain unwanted airspeed. Even if you have lots of altitude to work with, the airspeed gained by trying to half loop out of the inverted position can be extremely dangerous.

Recently, a pilot practicing for an aerobatic competition executed a maneuver called a split-S, which is a half roll to inverted, followed by a half loop back to level flight. Figure 5-12 depicts what a split-S looks like. Done correctly, the split-S is a very safe maneuver. But in this particular situation, the pilot started the maneuver at too high an airspeed, with cruise power on the engine. The result was that during the half loop the plane gained so much airspeed the wings were folded back on the plane and a crash followed.

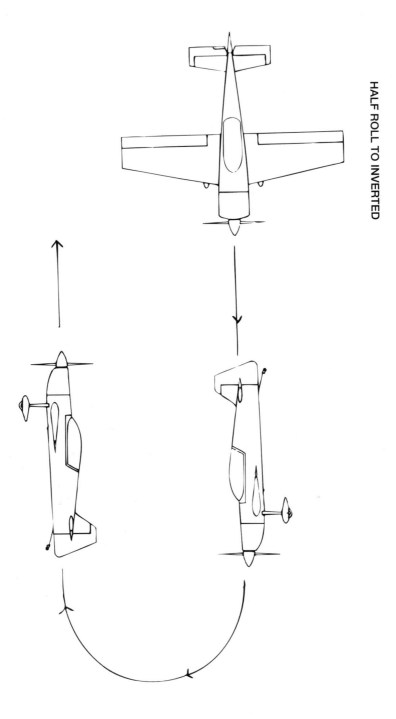

HALF ROLL TO INVERTED

Fig. 5-12 Split-S.

Be sure not to gain excessive airspeed by diving the airplane during recovery. Figure 5-13a shows how a plane should be rolled to upright from the inverted attitude after the spin stops. Figure 5-13b illustrates how the same plane does not successfully complete the half loop maneuver in an attempt to get back to level flight due to excessive altitude loss.

SUMMARY

Two very interesting spins are covered in this chapter, both of which most pilots have not heard of, let alone practiced. Unless you seek out specialized spin training, most flight schools do not have aircraft that can safely perform these spins. You should, however, be aware that you can get into these spins if you do not follow the proper spin recovery procedures. You need to know the spin recovery procedures for the plane you practice spins in, and follow them in the correct order to successfully recover from spins.

The first section of the chapter covered accelerated spins, discussing the pilot actions that can cause accelerated spins. Pushing forward on the elevator control while still holding pro-spin rudder does not break the stall. Instead, the change in angle of attack causes the spin's rate of rotation to increase, in some situations by a dramatic amount. This can cause pilot disorientation and excessive structural stress on the aircraft. Both of these factors can make it difficult to recover from the accelerated spin.

Accelerated spins are a precursor to crossover spins and a continuation of poor spin recovery technique. After entering an accelerated spin with the elevator control forward, inputting counter-spin rudder can cause the plane to cross over into an inverted spin. The Muller/Beggs recovery technique can be used to recover from this situation, with the same cautions about recovery success. Pilot disorientation is possible. Even more than in an accelerated spin, the crossover spin, with its rapid change from positive to negative g's, can cause the pilot to lose orientation. A half roll to upright is the quickest and safest way to recover from inverted flight back to straight and level.

If the plane you practice spins in has a different set of spin recovery procedures than the FAA-recommended procedures, it may also have different steps that will put it into an accelerated or crossover spin.

Always monitor your altitude during any spin practice. Too often, students become so focused on the spin, they fail to notice how much altitude the plane has lost. If having sufficient altitude to recover ever becomes a question, recover immediately. Wasting altitude in an attempt to complete a practice spin can lead to a potentially unsafe situation. By recovering early, you can always climb to a higher altitude and attempt the spin again.

To recover a plane from accelerated spins, flat spins, and other exotic spins, you need to get the plane back into a normal mode spin, whether it is upright or inverted. You can then recover from the normal mode spin using more conventional spin recovery techniques.

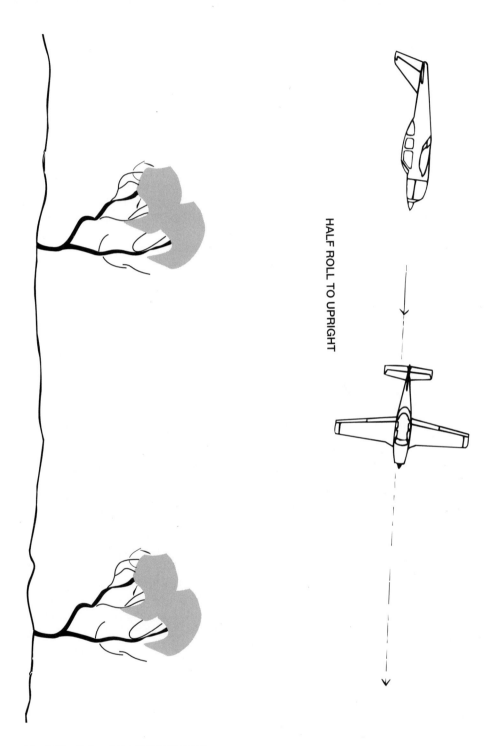

HALF ROLL TO UPRIGHT

Fig. 5-13a Roll upright after inverted spin.

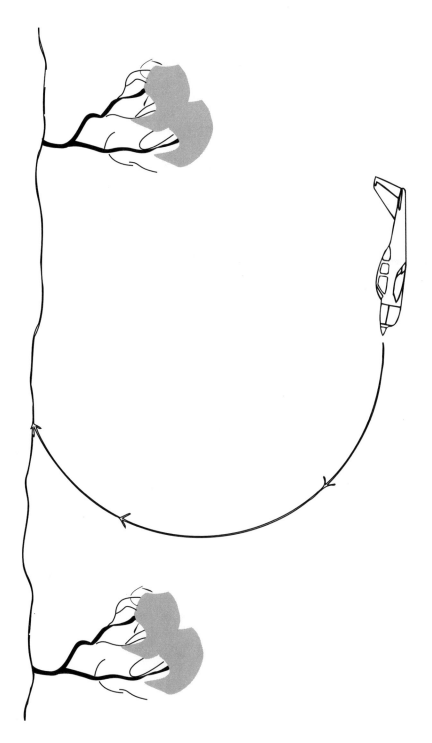

Fig. 5-13b Failed half-loop after inverted spin.

The problem lies in getting the plane back into the normal mode spin. The remainder of the book discusses how to use the Muller/Beggs spin recovery techniques to recover from various spins. For the Pitts S-2Bs I fly, this has the effect of transitioning the plane from an accelerated or flat spin into a normal mode spin, then recovering completely from the spin.

6

Inverted Spins

I KNOW I AM IN A MINORITY, BUT I LIKE DOING INVERTED SPINS. IN addition to the sheer fun of executing them, inverted spins help you learn to deal with a number of factors you are not exposed to during normal spins. Proper use of controls in very unusual attitudes, recognition of aircraft attitude with reference to the horizon in unusual attitudes, and a sharpening of spin recognition and avoidance skills are but a few factors that play into inverted spins. Chapter 5 touched briefly on inverted spins as part of the crossover spin, and this chapter provides an in-depth look at these enjoyable spins.

The discussion begins with inverted flight, followed by an explanation of inverted stalls. These two topics lay a solid foundation for a discussion of inverted spins. Factors affecting the pilot during inverted spins are also reviewed. Later sections describe inverted spin entry and recovery. For these topics, proper use of controls, disorientation during the spin, and spin recovery are each discussed. The photographs included document the forward and side views as you would see them from the pilot's seat during the maneuvers. While they cannot capture the exhilaration of doing these spins, they can give you an idea of the attitudes the plane passes through at the various stages of the spin.

In a correctly certified plane and with the proper preparation, inverted spins are a lot of fun. As with all spins covered in this book, be sure they are performed with a properly qualified instructor and in the right plane. Because the plane must be inverted to begin an inverted spin, you need to wear a parachute. You should also have an airplane equipped with inverted fuel and oil systems.

INVERTED FLIGHT

In my opinion, inverted flight is the best way to fly an aerobatic airplane. It gives you a whole new perspective of the world. There's something about executing a half roll to inverted, then looking up to see the ground passing below you. It typically takes time for a new aerobatic pilot to get used to the different view that this flight attitude brings, however. For most of my aerobatic students, we usually attempt short periods of inverted flight around the third or fourth lesson. At first they stare intently ahead, focusing directly in front of the plane. I ask them to look to the left and right after a little bit, so that they see what the entire span of vision is while the plane is upside down. The first exposures are short, limited fifteen to thirty seconds, depending on how the pilot reacts.

Gradually new pilots begin to accept the idea of being upside down for extended periods of time, and soon transition to longer periods of inverted flight. They begin to recognize climbs, dives, and banks and are able to fly the plane in the same manner they do when the plane is right side up.

Physical Aspects of Inverted Flight

Inverted flight brings new sensations for those who try it. First is the view they get of the world from the plane. Most airplanes capable of inverted flight have better visibility forward and above the plane, as opposed to below it. The wings and fuselage typically block your view down to some extent. Yet when the plane is upside down, forward and up are (in relation to the plane) now giving you a view of the ground. The sky above you is probably blocked by the airplane's fuselage and wings. Some aerobatic aircraft have clear plastic in the bottom of the plane that may offer some view to the pilot above the airplane, but this is typically very limited in its field of view. Pilots must adapt to the new perspective that the inverted flight attitude provides. Their scan of airspace needs to change to accommodate looking for other aircraft while inverted. (Yes, you should still look for other planes when you fly upside down, in addition to checking engine gauges and other instruments.)

Depending on the plane you are flying and the configuration of the wings, you will find your field of vision can be affected by the wings as well. For instance, if you fly a Decathlon, which is a high-wing plane, in upright flight the wings are above you, partially blocking the view of the sky above. But when you fly inverted, the wings are now below you, blocking the view of the ground. Now the sky above you is unrestricted, except by the wing struts. Low-wing aerobatic planes are the reverse of a high-wing, blocking your field of vision of the sky above you when the plane is inverted. Biplanes block your vision above and below to some extent, whether you are upright or inverted, so you should take special care in clearing the area you are flying in before performing any maneuver in a biplane.

Until pilots become accustomed to the new viewpoint offered by inverted flying, picking out the horizon is more difficult, especially when the pilot is looking to the sides. All of the normal visual queues are altered and require that the pilots re-acquaint themselves with the old visual flight references in this new perspective of the world of flight. In some cases, the pilots must change where they have been looking to pick up the correct attitude queues. Wings can sometimes block the normal views, so pilots must look at a different angle to see around them. Occasionally only smaller pieces of the horizon are available for attitude reference due to this restricted vision, and the pilot must adapt accordingly to find the correct reference points.

The most striking physical sensation for pilots the first time they fly inverted is usually that their seatbelt is not tight enough. Many first-time inverted pilots drop an inch or two as the plane goes inverted. They hang in the seatbelt, a fair amount of space between them and the seat. If you are going to fly inverted, be sure to cinch the lap belt down as tight as possible without causing yourself physical discomfort. The lap belt should take the majority of your weight during inverted flight, not the shoulder harnesses. Some pilots snug the shoulder harnesses down very tight, feeling this will better secure them in the plane. However, you will still need to move forward and turn your shoulders when you are upside down, and excessively tight shoulder harnesses will restrict your freedom of movement. Pilots seem more comfortable with inverted flight when they do not feel like they're floating off the seat, so tighten the lap belt as much as you can.

Once you are firmly secured to the seat, the next physical sensation you will probably notice as you cruise along inverted is a certain amount of pressure in your head. When you fly upright in unaccelerated straight and level flight, you experience a positive one g load on your body. This is normal and you are accustomed to it. When the plane is flying inverted in unaccelerated straight and level flight, you are subjected to a negative one g load. This causes more blood to pool in your head, which is now the low point of your body, resulting in pressure similar to what you might feel if you hung by your feet.

Most people are able to adapt quickly to this particular sensation, with the pressure becoming second nature. You may also experience sinus pressure if you have any congestion in your head. If you find that the pressure becomes too distracting, wait until the sinus infection clears before further inverted flight. For those who fly more advanced aerobatic maneuvers, negative g's become a part of every flight. But it takes time to build a tolerance to negative g's. Start out slowly, and do not attempt to load yourself up too quickly with negative g's. Depending on the plane you fly, you may experience two or three negative g's during an inverted spin. This is typically not a problem for pilots, but keep the initial amount of time spent inverted short to avoid overloading yourself.

Once you are upside down, you will also need to make an effort to keep your feet on the rudder pedals. While you are right side up, gravity keeps your feet in place against the pedals; but gravity works against you after you roll inverted. With a little practice, pilots become accustomed to the amount of effort necessary to hold their feet and legs in the correct position during inverted flight.

Be careful not to brace your feet against the rudder pedals in an attempt to hold your feet in place. This will reduce your feel of the rudder pedals during inverted flight and make it more difficult for you to use the correct amount of rudder during maneuvers. It is also more fatiguing, quickly tiring your legs. Unless you are a hardcore, negative g addict, you can easily adapt to the negative g's you experience during straight and level inverted flight and inverted spins. Some of the more advanced negative-g aerobatic maneuvers require more time and conditioning.

Aerodynamics and Inverted Flight

Will an airplane not equipped with a symmetrical airfoil fly in inverted flight? How does the symmetrical airfoil used on many aerobatic aircraft create lift if it has the same shape above and below the cord line? How will dihedral, used to make the plane more stable during normal flight, affect the plane once it is inverted? These and other aerodynamic factors are discussed in this section of the chapter. Some pilots are skeptical that a plane really can fly in a controlled manner upside down. The purpose of this section is to assure you that properly designed planes are capable of flying for extended periods, certain limitations notwithstanding, in an inverted attitude.

Figure 6-1 illustrates a typical, nonsymmetrical Clark Y airfoil. During normal flight lift is produced by two methods, pressure differential and impact pressure. In a plane equipped with the conventional airfoil shown in the figure, air flowing over the top of the wing must travel a longer distance than the air that passes under the wing. The air moving across the wing's upper surface travels faster than the air flowing under the wing. This higher speed causes a reduction in pressure along the wing's upper surface, and a pressure differential between the upper and lower surfaces of the wing. This pressure differential generates part of the lift that allows the airplane to fly.

Air impacting the underside of the wing also generates lift. The Newtonian physical law, "for every action there is an equal and opposite reaction," applies in this particular case. Figure 6-2 is repeated from Chapter 3, because it applies to inverted flight aerodynamics. The lower portion of the figure illustrates how air is deflected downward, which imparts an upward motion to the wing. As a result, the wing generates additional lift, which helps the plane get into the air. This is a simplistic approach to a complex physical event, but it provides the basics for the discussions that follow. The point is to understand how a conventional airfoil generates lift under normal flight through pressure differential and impact pressure.

Can this same airfoil be used in inverted flight situations? The airfoil is no longer being used in the manner for which it was designed. The curved surface

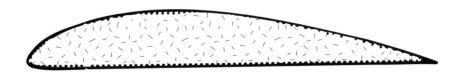

Fig. 6-1 Clark Y airfoil.

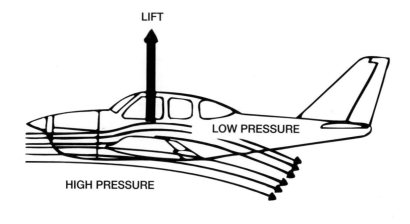

LIFT

LOW PRESSURE

HIGH PRESSURE

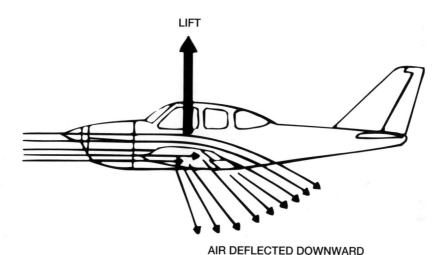

LIFT

AIR DEFLECTED DOWNWARD

Fig. 6-2 Pressure/deflection lift.

is now the lower portion of the wing, and the shorter distance for air to travel is now on the wing's upper surface.

Figure 6-3 is a pictorial definition of the angle of incidence. The angle of incidence is the angle formed between the cord line of the wing and the longitudinal axis of the plane. Angle of incidence is normally designed into an airplane to provide optimum flight characteristics at cruise flight. Even though the plane may appear to be at a level pitch attitude, the wings of most general aviation aircraft will have a positive angle of attack as a result of this angle of incidence. The amount of the angle varies depending on aircraft design. This is an important concept to remember: even when the plane is level, the wings will probably have a positive angle of attack, which is used to help generate lift.

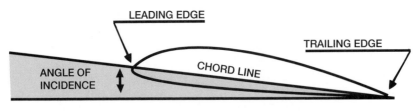

Fig. 6-3 Angle of incidence.

Now take the Clark Y airfoil and put it in an inverted flight situation. If you put it at a 0° angle of attack, the air passing under the wing has the longer distance to travel, so you lose some of the lift that was generated when it was right side up. But if you increase the angle of attack, you can cause the airflow over the top of the wing to take a longer course than the air flowing along the underside of the wing. The additional angle of attack also increases the amount of lift produced by impact air. Even though the curved surface of the wing is now on the underside, the air impacting this area of the wing also produces lift.

This increase in the angle of attack now begins to generate lift, even though you are not using the airfoil in the manner for which it was originally designed. But because the plane has a positive angle of incidence when it is right side up, the wing has a negative angle of incidence when the plane's longitudinal axis is level in inverted flight. Therefore, to achieve the angle of attack you need to generate lift, you must pitch the nose of the plane up with forward elevator pressure when you are inverted. Figure 6-4 shows the plane with the inverted pitch angle needed to generate lift. The higher angle of attack will also generate additional impact lift in addition to pressure differential lift.

I have had the opportunity to fly aerobatics in a Citabria equipped with inverted fuel and oil systems. The Citabria's wing is the classic flat underside, curved upper surface airfoil design. Flying that plane upside down was exciting, but it took a large amount of nose up pitch attitude when the plane was inverted to maintain altitude. In fact, there is a very small range of pitch between which the plane is able to maintain altitude and above which the plane will stall. Flying that plane was a great experience, and it showed me that under the right circumstances a normal airfoil can keep a plane flying inverted.

In an emergency situation, such as a roll to inverted induced by airliner vortexes, this means that many single-engine general aviation planes have a very good chance of getting back to upright level flight. Just because a plane ends up upside down, or in any other unusual attitude, does not automatically mean the aerodynamics that keep the plane flying in normal flight quit working. There are many factors that will determine the successful outcome of one of these emergency situations. For this reason, do not quit flying the airplane just because it ends up inverted. Also for this reason, you can consider

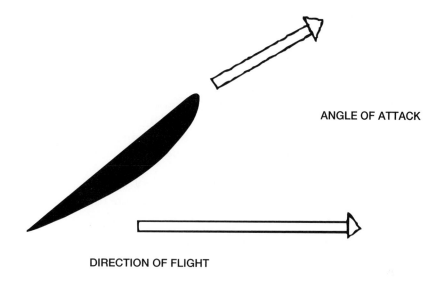

ANGLE OF ATTACK

DIRECTION OF FLIGHT

Fig. 6-4 Inverted pitch angle (flat bottom airfoil).

aerobatic training as a way to enhance safety because you gain experience in dealing with unusual attitudes.

To answer another question posed at the beginning of this section, how does the dihedral of the Citabria wing affect the airplane when it is inverted? When the plane is upright, dihedral is designed to help keep an airplane's wings level. Figure 6-5 shows how dihedral causes more lift to be generated by the lower wing whenever a plane's wings are upset from level flight. As a result of dihedral, additional lift causes the lower wing to generate more lift and rise, leveling the wings. When you fly the Citabria during inverted flight, the plane really wants to roll off in one direction or another. This negative dihedral, or anhedral, causes less stability when the plane is inverted. You must stay on top of the roll tendency of the plane in order to maintain wings-level flight. Here is yet another reason to keep flying a plane, even in unusual attitudes.

DIHEDRAL ANGLE

Fig. 6-5 Dihedral.

Symmetrical Airfoils

This section covers an airfoil type used on many aerobatic airplanes, which is the symmetrical airfoil. As its name implies, the symmetrical airfoil, pictured in Figure 6-6, is designed with the same shape on the upper and lower surfaces of the airfoil. Aerobatic aircraft designers use variations on the basic symmetrical design, altering the thickness of the airfoil and other physical characteristics of the design. Both the upper and lower surfaces of the airfoil in Figure 6-6 resemble the upper surface of the normal airfoil depicted in Figure 6-1. Because the upper and lower surfaces have the same shape, how does this airfoil generate lift?

The air flowing along either the upper or lower portion of the wing has the same distance to travel, which will prevent the pressure difference above and below the wing that is seen in normal airfoils. The answer is that when the plane is upright, the angle of incidence causes the wing to have an angle of attack that creates lift. The positive angle of attack causes the distance over the top of the wing to become longer than the distance along its lower surface. The wing is now able to produce lift as a result of pressure differential. Impact lift is also created as a result of the angle of attack of the wing. Like our Citabria example, even though a curved surface is on the underside of the wing, lift is generated by air impacting it. Because of the plane's design, it will be in level flight while the wings have a positive angle of attack.

Using the same principles as discussed for normal airfoils, an angle of attack must be established when the symmetrical airfoil is inverted to provide the lift necessary to keep the plane flying. Unlike the normal airfoil, the symmetrical airfoil generates lift equally as well when it is upside down as long as the correct angle of attack is maintained. You create this angle of attack by using forward pressure on the elevator control as the airplane becomes inverted. Figure 6-7 shows an imaginary airplane with a symmetrical airfoil flying in both the inverted position and in upright flight. Notice the angle of attack in both positions is the same, but the pitch angle of the nose above the horizon is greater in the inverted position. From the airfoil's perspective, whether the plane is upright or inverted does not matter; it can generate lift in either position.

Fig. 6-6 Symmetrical airfoil.

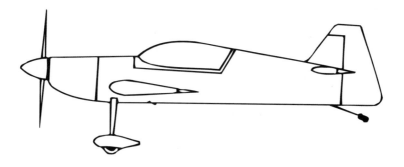

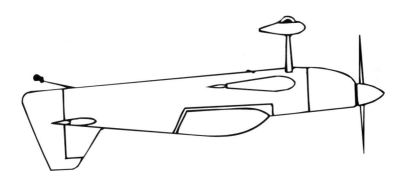

Fig. 6-7 Symmetrical airfoil plane upright/inverted.

Flight Control Use During Inverted Flight

Many first-time inverted flight students are surprised to find that most planes designed for inverted flight are well behaved when they are inverted. The flight controls typically provide the same level of aircraft control when the plane is inverted as in upright flight. All that is needed is for the pilot to adjust his or her reactions to this new but very enjoyable flight regime.

This section briefly discusses the control inputs necessary to accomplish basic flight maneuvers while the plane is inverted. Straight-and-level flight, climbs, descents, and turns are explained. The purpose of this section is to familiarize you with the concepts of control use during inverted flight. Practicing inverted flight in the proper aircraft, with a qualified flight instructor, is the best way to learn how to fly inverted.

Inverted straight-and-level flight is relatively easy for a pilot to master. I have my students initiate inverted flight by doing a half roll, putting the plane in an inverted position. During the first moments of inverted flight, most students hold the stick rigidly in their grip once the plane is upside down. When the half roll is properly executed, as the plane approaches inverted, the pilot should push forward on the elevator control, putting the plane at the correct pitch attitude to maintain level flight. In many cases, student pilots do not give an adequate forward push on the elevator control, allowing the nose to drop and the plane to descend.

Pilots may also have some trouble recognizing the horizon and determining if the wings are level. This difficulty is normal and quickly overcome as the pilot begins to adapt to inverted flight. Once the pilot understands where to hold the control stick, as long as the air is smooth the plane flies along straight and level with very little control correction needed. After they get through these initial moments, new inverted pilots begin to perceive some of the control pressures necessary to maintain level flight. The comment I have heard most from these initial forays into inverted flight is how much forward elevator pressure is necessary to hold the nose at the correct attitude.

When the plane is inverted, pushing forward on the control stick raises the nose of the plane, while pulling back on the stick lowers the nose. I have students trim the elevator to help them maintain manageable elevator control pressures when the plane is inverted for extended periods of time. When the plane is inverted, move the elevator trim toward the nose down marking to help raise the nose. If you do trim the plane when you are inverted, remember that the plane will have nose down tendencies when you return to upright flight and you will need to compensate for this until you re-trim the plane.

Pilots seem to master the basic pitch concepts rapidly, learning how to use the elevator control to put the nose where it needs to be to maintain level flight. Figure 6-8a shows how pushing forward on the control stick causes the plane's nose to rise. Figure 6-8b shows the nose of the plane dropping when the pilot pulls the stick back. Note the position of the elevator in each case. Another way to view elevator use is as follows. Whether upright or inverted, if you pull the stick toward you, the nose moves toward you. If you push the control stick away from you, the nose moves away from you.

Ailerons work in exactly the same manner whether the plane is upright or inverted. To roll to your left, push the control stick to your left. To roll to your right, push the control stick to your right. Some pilots are initially confused, thinking that the ailerons are reversed when the plane is inverted. This is not the case. Figure 6-9 shows a plane in the inverted position, with the stick to the

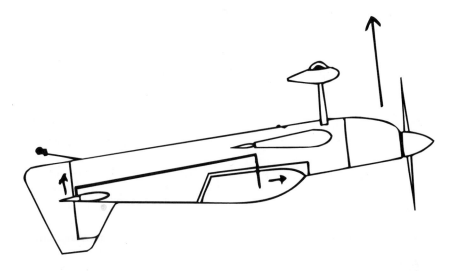

Fig. 6-8a Forward stick raises nose of inverted plane.

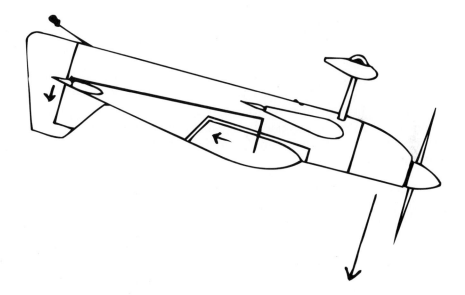

Fig. 6-8b Aft stick lowers nose of inverted plane.

left. The ailerons are displaced in the normal manner, causing the plane to roll to the left. Turns can be initiated from inverted flight as well, although there is one slight complication. When a plane is inverted, to make a coordinated turn, you use rudder OPPOSITE the aileron input.

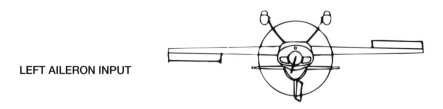

LEFT AILERON INPUT

RIGHT RUDDER INPUT

Fig. 6-9 Inverted left aileron input.

Figure 6-10 depicts why this is necessary. This is a common example, used in many aerobatic texts. Imagine two pilots one inside the cockpit, the other on the belly. Briefly, the pilot sitting on the belly of the plane, who is right side up, is using extensions of the same flight controls as the pilot inside the plane, who is upside down. Now, to roll to the left, the pilot on the belly of the plane inputs left aileron and left rudder. To the pilot inside the plane, the stick moves to the left, but the rudder input is the right rudder. The plane does not care that it is upside down; it just needs the rudder to counteract the adverse yaw caused by the lowered aileron. For this reason, the pilot inside the plane needs to use rudder opposite the aileron input.

While confusing at first, pilots quickly become accustomed to using the correct control inputs to bank a plane when it is inverted. Use of the controls during inverted flight is straightforward, requiring only that you make a few minor adjustments to fly the plane in a coordinated manner. Pilots are able to make inverted climbs, descents, and turns, with the plane flying in the same manner as during upright flight.

Inverted Flight and the Engine

If you intend to practice inverted flight, you need to fly a plane with an engine equipped to run when the plane is upside down. Special hardware must be added to the engine for this purpose. For instance, most fuel tanks have the fuel pick-up port at the lowest point in the tank. This works fine as long as the plane is upright. However, when you turn the plane upside down and do not maintain positive g's, the fuel moves away from the port, allowing air to enter the fuel line, and eventually resulting in engine stoppage and a very quiet cockpit.

I have flown aerobatics in a Cessna 152 Aerobat, an aerobatic version of the 152. This particular plane was not equipped with inverted fuel or oil systems. Every time I did a two-point roll, during which the plane briefly pauses while inverted, the engine quit due to fuel starvation. Once I rolled upright, the engine would start again, but this type of power loss does not lend itself to

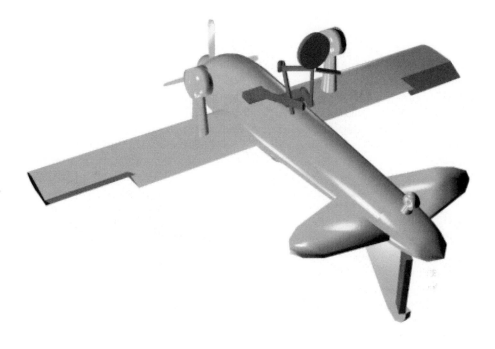

Fig. 6-10 Inverted plane in left turn.

extended inverted flight time. In order to overcome this problem, fuel systems have been built that allow fuel to flow to the engine while the plane is inverted. This section describes a typical inverted fuel system design.

Once you have dealt with the need to keep fuel flowing to the engine, oil pressure becomes the next consideration. Most engines have the oil pump pickup located at the bottom of the oil pan. The pump takes oil in at this point, and then pumps it through the engine. When an airplane is upside down, all of the oil in the oil pan shifts to what used to be the top of the crankcase. As a result, the oil pump pickup will not have sufficient oil available to lubricate the engine. The worst-case result in this scenario would be an engine seizure due to increased friction and heat. Even at best, the loss of oil pressure will increase engine wear and reduce the life of the engine. To overcome this problem, inverted oil systems have been developed to allow oil to continue to flow through the engine when it is inverted. A typical inverted oil system is also covered in this section.

Inverted Fuel Systems

Figure 6-11 depicts the first component of a typical aerobatic aircraft inverted fuel system, the fuel tank. When a plane is upside down, or pulling negative g's for that matter, the intake port in the tank can "unport." This means that fuel no longer covers the port, and the flow of fuel is interrupted to the engine.

To overcome this problem, aircraft designers came up with a very simple solution. A tube, called a flop tube, is attached to the intake port inside the fuel tank. As the name implies, this tube flops around the fuel tank, following the motion of the fuel as the plane flies. A screen or filter of some type is typically attached to the pickup end of the flop tube to help prevent foreign matter from entering the fuel line. When the plane is upside down, or pulling high negative g's, the flop tube stays with the fuel as it moves, maintaining the flow of fuel to the engine.

Inverted fuel systems also typically have a fuel pump incorporated into them. Many high-wing aircraft such as the Cessna 152 Aerobat use gravity feed systems. In this design, when the fuel is below the engine, such as when the plane is inverted, fuel no longer flows from the tank to the engine. Even with a flop tube, a gravity-fed system will not provide fuel to the engine during inverted flight, because gravity no longer pulls the fuel down from the tanks to the engine.

The final piece of the fuel system configuration is a fuel injection system. Most carburated fuel systems are designed to provide fuel to the engine under positive g conditions. Float systems in the carburetor will typically not work correctly when the plane is in an inverted attitude or a high negative g maneuver. Injected engines bypass the carburetor, injecting fuel directly into the

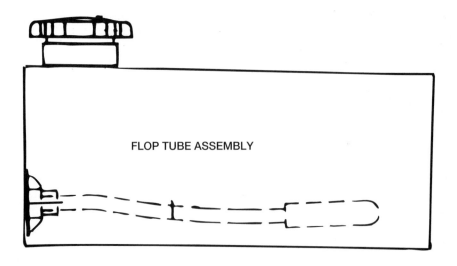

FLOP TUBE ASSEMBLY

Fig. 6-11 Inverted fuel tank.

engine's intake valve port. Because the fuel pump provides a positive fuel flow at all times, the injected system will continue to receive fuel even when the plane is inverted. Carburated engines are dependent on the carburetor correctly metering fuel to the engine. To do so, the needle and float inside the carburetor must remain at the correct level, controlling the amount of fuel flowing into the carburetor throat. When the plane is inverted, the float no longer operates in the correct manner and fuel no longer flows into the engine at the correct rate.

INVERTED STALLS

This section reviews a maneuver that is necessary to enter an inverted spin, which is the inverted stall. This is performed in exactly the same manner as a normal stall, but the plane is inverted instead of upright and elevator use must be modified accordingly. You must take all of the normal pre-stall precautions, such as adding carburetor heat (if required) and executing clearing turns prior to setting up for the stall. To initiate the inverted stall, you make a half roll to inverted, putting the plane in an inverted attitude. Figure 6-12a shows the forward view of the plane in this inverted, level flight attitude. The view to the left side of the plane in Figure 6-12b shows the angle of attack that is necessary to maintain altitude.

As with upright spins, I like to begin inverted spins on a cardinal heading, which makes it easier to track the number of turns the plane has completed. Power is eased back to idle, with forward elevator control added to maintain

Fig. 6-12a Inverted level flight: forward view.

Fig. 6-12b Inverted level flight: left side view.

altitude as the plane slows. Figure 6-13a shows the plane's attitude just prior to a stall. This forward view shows the nose well above the horizon, but not an extreme amount. Figure 6-13b shows this same point just prior to the stall as viewed to the left. It is not necessary to force the nose to a very high pitch angle for inverted stalls. Some pilots get in the habit of pulling the nose well above the horizon during upright stalls and attempt to use the same technique during inverted stalls. Instead, a nice, smooth forward elevator pressure will help maintain altitude while the plane is inverted and reduce the airspeed at a comfortable rate.

As the plane stalls, it demonstrates stall characteristics similar to those in an upright stall. The plane will probably exhibit a buffet prior to the stall, with the nose dropping or breaking as the plane stalls. When this takes place, you typically only need to reduce the angle of attack by releasing some of the forward elevator pressure. This reduction in pressure should be enough to allow the nose to drop just below the horizon, much as was demonstrated during stalls covered in Chapter 2.

First-time inverted stall pilots have a tendency to let the nose fall too far below the horizon, putting the plane into an inverted dive. Figures 6-14a and 6-14b show the forward and left view after the plane has recovered from the stall. You can compare these views to those of upright stalls in Chapter 3. Every plane is different; but for most, it is not necessary to dive excessively during inverted stall recovery. I like to have my students experiment with both power-

Fig. 6-13a Inverted stall entry: forward view.

Fig. 6-13b Inverted stall entry: left side view.

Fig. 6-14a Inverted stall recovery: forward view.

Fig. 6-14b Inverted stall recovery: left side view.

off and power-on recoveries from inverted stalls, to give them an appreciation for the difference in how the plane recovers. When making power-on recoveries, add power gently as you recover and be prepared for any pitch changes that the additional thrust may induce. Until you become adept at attitude control while inverted, adding power during stall recovery may induce excessive pitch changes. Because safety dictates that inverted stalls be done at a safe altitude, leaving the power off during stall recovery and adding power once airspeed has increased and the nose raised to level flight should pose no safety problem.

You should use rudder to keep the wings level during the stall, as you do with upright stalls. Excessive use of ailerons will deepen the stall on one wing and may put the plane into an unintentional spin. I recently attended a flight instructor refresher seminar, during which the instructor showed a video taken by a camera crew from a boat on a lake. The scene showed a single-engine plane that was obviously in trouble. The aircraft was losing altitude. The pilot was holding the plane on the edge of a stall as he tried to clear trees at the lake's edge, for a water landing in the lake. Watching the video several times made it clear the pilot was doing his best to avoid stalling before impacting the water. First one wing would drop, and then the other, with the plane on the ragged edge of a stall. The pilot was using rudder to keep the wings as level as possible while he fought to control the plane.

The instructor commented that the pilot in this video had received good flight instruction related to stalls. Instead of trying to hold the wings level with ailerons, which might have put the plane into a spin, the pilot used rudder to level the wings. As a result of this technique, the plane impacted the water and remained upright. The pilot and his passengers were injured, but they lived through the crash. I have always taught my students to use rudder to keep the wings level during stalls, but to witness this video drove the point home.

Once you have recovered from a stall, do not force the nose back to level flight before the airspeed has increased sufficiently. If you try to do this, you may cause a secondary stall. Use rudder to counteract P-factor, slipstream, and the other turning tendencies. From the perspective of the pilot sitting on the belly of the plane in the previous example, the plane is still trying to turn left as the angle of attack increases, requiring that right rudder be used to counteract this tendency. From your perspective inside the cockpit, you push on left rudder to maintain coordinated flight. Some aerobatic planes actually have two balls for determining slips and skids. One is right side up and the other is upside down. When the plane is inverted, the pilot can tell when the plane is not being flown in a coordinated manner by the position of the ball.

Common Errors

The most common errors I see during inverted stalls are incorrect use of the elevator during both stall entry and recovery. Pilots sometimes do not push forward enough on the elevator control to initiate a stall once power has been reduced, and the plane merely enters an inverted descent. At the other end of the spectrum, some pilots give a nice negative 3-g push on the elevator control stick as they pull power back. This helps the plane reach a pitch angle of 45° or greater before the pilot realizes that slightly too much forward pressure has been applied. The nose of the plane can drop sharply when stalled from this attitude. Pilots also tend to not recover from the inverted stall because they do not want to drop the nose below the horizon, or they allow the nose to drop to an angle that is much greater than necessary. In addition to these errors, the other common errors discussed in Chapter 2 can apply to inverted stalls as well.

INVERTED SPINS

Hanging against your safety harness as the plane executes an inverted spin brings you a whole new perspective of the world! I have to admit that I'm a negative-g addict, and inverted spins are one way I satisfy that need.

Inverted spins are very similar to upright spins, from spin entry to recovery. Your attitude (both physical and mental) are different, however. Inverted spins are part of the aerobatic maneuvers a pilot must perform in the more advanced categories of competition. They require that the pilot have a firm understanding of spin recovery procedures. For pilots accustomed to upright spins, the opposite use of elevator controls during inverted spin entry and recovery can require a spin or two before the pilots become comfortable with them. But as a result of this learning experience, the pilots become even more comfortable with spins. They begin to realize that spins are not uncontrolled situations when done correctly, in properly certified airplanes. The aerodynamics of spins also begin to make sense as they see that a plane behaves in the same manner in both upright and inverted spin situations.

Inverted Spin Entry

Inverted spins begin with an inverted stall. As the plane stalls, you give firm rudder input in the direction you want to spin (from your perspective inside the cockpit). In other words, if you want to spin to your left, push left rudder. If you want to spin to your right, push right rudder. Do not become confused in thinking you need to push opposite rudder to the direction you want to spin. As the plane begins to yaw and roll in the direction of the rudder input, ease the elevator control against the forward stop to ensure that the plane remains stalled as it enters the spin. Recall that this forward elevator input will increase the angle of attack, just as pulling the stick fully aft did in an upright spin. Keep the control stick centered as you push it against the forward stop.

Figures 6-15a and 6-15b show the forward and side views as the plane begins to enter the spin. As you can see in this example of a spin to the left, the plane banks as the nose drops. The plane's progress into the spin is documented in Figures 6-16a and 6-16b. Here you see the pitch and bank angles of the plane.

Fig. 6-15a Inverted spin entry: forward view.

Fig. 6-15b Inverted spin entry: left side view.

Fig. 6-16a Inverted spin: forward view.

Figure 6-16b Inverted spin: left side view.

When you practice them, you will find that inverted spins have a higher rate of rotation than upright spins. As a result, the plane will seem to accelerate into the spin more quickly than its upright counterpart. As the spin progresses, you will probably experience increasing negative g forces, until they equalize to some degree. Depending on the overall spin characteristics of the plane you are flying, it may oscillate in its pitch attitude and rate of rotation as the spin progresses, as with upright spins. Try to pick up these changes in the plane's attitude and roll as it continues in the spin. Keep neutral ailerons as you hold the control stick and maintain full rudder input in the direction of the spin while in the spin.

Recovering from Inverted Spins

This section reviews two methods for recovering from inverted spins. The first set of recovery procedures you might use to recover from an inverted spin is the Muller/Beggs procedure. I teach students to use this recovery technique when they first begin to learn inverted spins. To review briefly, the procedures are:

1. Power off.
2. Let go of the control stick.
3. Input full rudder opposite the direction of the spin.
4. Neutralize rudder when rotation stops and recover from the dive.

This procedure makes it very easy for the new inverted spin pilot to recover from the spin. You do not need to know where to position the control stick. As long as you pull the power off, let go of the stick, and then push full opposite rudder, you will recover from the spin. The first few times pilots do inverted spins, they typically become disoriented due to the number of new sensations. Using the Muller/Beggs technique for recovery reduces the amount of information they must work with and results in a successful spin recovery. Figures 6-17a and 6-17b show the forward and side view during recovery.

Once students become more accustomed to inverted spin, we begin to work on active spin recovery control techniques. Especially for aerobatic competition, this active spin recovery is necessary to recover the plane from the spin on an exact heading. The procedures for this type of recovery are:

1. Power off.
2. Input full rudder opposite the direction of the spin.
3. Release forward elevator control pressure.
4. Neutralize rudder and recover from the dive when rotation stops.

These procedures may look familiar; they are the same ones you use for upright spin recovery, except for how the elevator control is used. Instead of releasing backpressure to break the stall, you release the forward pressure used during an inverted stall/spin. If you are attempting to come out of the spin on a particular heading, you lead the spin recovery point with counter-spin rudder input. Depending on the plane and its rate of rotation, anywhere between 90° and 45° prior to the desired recovery heading, you input full rudder opposite the spin's direction of rotation. Then, just prior to the recovery point, release

Fig. 6-17a Inverted spin recovery: forward view.

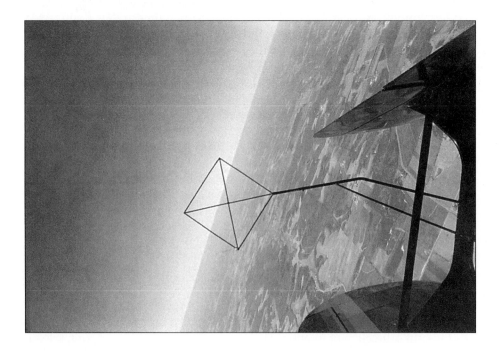

Fig. 6-17b Inverted spin recovery: left side view.

the forward pressure on the elevator, stopping the aircraft on the correct heading. Each plane behaves differently during inverted spin recovery and you need to experiment with the plane you fly to find the timing of control inputs that works best for stopping on a point. During competition, inverted spins are still in the incipient phase during recovery, which aids in being able to stop a spin on point. If you enter the developed portion of a spin, it may take longer to recover. Remember to center the rudder pedals after the spin has stopped. The airplane you fly may require a different set of recovery procedures than either of those discussed in this section. If this is the case, use those procedures. You should always read, understand, and follow the spin recovery procedures recommended by the aircraft's manufacturer.

Improper Inverted Spin Techniques

All the improper spin entry techniques previously discussed apply to inverted spins. The most common error I have noticed is students not using enough forward elevator to get the plane into a good solid stall before attempting to enter the spin. Especially when new to inverted flight, pilots have a tendency to release the forward pressure on the elevator control, allowing the nose of the plane to drop. This can cause the plane to recover from the stall/spin and enter a dive or inverted spiral.

Another common error during inverted spin entry is the lack of full rudder input as the plane stalls. Given the number of new sensations pilots are subjected to during their first few inverted spins, they will concentrate too closely on one aspect of the spin, such as maintaining full forward elevator as the plane stalls, and forget to use full rudder input. This is not uncommon and requires only practice before the pilot automatically uses full rudder inputs.

Once pilots get the hang of getting into an inverted spin, they will occasionally fail to use full opposite rudder to recover. As has been mentioned before, always use full counter-spin rudder during spin recovery. If you get into the habit of using full rudder inputs, you will not be surprised, as some pilots are, when use of only partial rudder during spin recovery fails to stop the plane's rotation. Occasionally pilots have not secured themselves tightly enough with the safety harness, so they float several inches off the seat cushion when they are inverted. Depending on how long their legs are, the pilots may not be able to fully extend the rudder pedal during recovery. This can be a problem, so be sure you strap yourself in as tightly as is comfortably possible. If you are unable to push the rudder pedals to the stops when you are sitting on the ground, adjust the seat to allow you to make those full inputs. In some cases, cushions may be necessary to accomplish the correct adjustments.

Disorientation is not uncommon for the new pilot during inverted spins. This can cause some hesitation when the pilot is uncertain what the control inputs should be for recovery. Delayed recovery and incorrect control inputs can result from this lack of familiarity with inverted spins. Like upright spins, after the first few spins have been completed, the pilot typically begins to have a clearer understanding of what the plane is doing and what control inputs should be used to recover from the inverted spin.

As with all spins, keep the ailerons neutral during spin entry, the developed spin, and spin recovery. Use of ailerons can produce a flat spin mode, or increase the rate of rotation. Both of these scenarios have been discussed in the last two chapters, but sometimes pilots try to stop the rotation with aileron instead of rudder. Unless the manufacturer of the plane you are flying recommends using ailerons, keep them neutral throughout the stall and spin.

Improper spin recovery sequence can also have the same types of effects on inverted spins as for upright spins. Accelerated inverted spins can establish a very rapid rate of rotation. Once in an accelerated inverted spin, I always use the Muller/Beggs spin recovery technique.

SUMMARY

I enjoy flying upside down. The process of learning how to fly upside down correctly makes better pilots, and it's an awful lot of fun. The first sections of the chapter discussed inverted flight and how it affects both you and the airplane. For a properly designed plane, inverted flight is no different than upright flight.

Inverted fuel and oil systems allow an engine to continue to run even during extended periods of inverted flight. Some manufacturers do place limitations on how long a plane is capable of inverted flight, though. Typically, the limits have to do with how long fuel will be available from the inverted fuel system when the plane is upside down. There may be other limiting factors, so it's a good idea to check the plane's operations manual to verify these limitations before you attempt inverted flight. How an airfoil generates lift during inverted flight was also reviewed. The intent was to provide you with a practical understanding of how the plane continues to fly while upside down.

How you might feel during inverted flight was also covered in this chapter. It does not take long to get used to, but the first few minutes of inverted flight place a new series of demands on the pilot. The view of the world changes dramatically when the plane is upside down. Pilots occasionally have trouble seeing the horizon clearly or determining the plane's pitch or bank angle. There is also a certain lack of confidence in how to use the controls to make the plane fly as it should. These feelings usually last only a short period of time and pilots quickly pick up the same cues they use during upright flight. Be sure you are strapped in tightly by your safety harness to reduce the gap that can form between you and the seat while you are upside down. Pilots seem to relax more quickly when they feel they are strapped in to the point that they are part of the plane, wearing it like a tight pair of blue jeans.

Inverted stalls were covered in this chapter as well. In reality, they are no different than upright stalls. The same aerodynamic forces act on the plane in both flight regimes. Stall entry and recovery are made using the same control inputs, adjusting for the fact that the plane is upside down.

The inverted stall review led into a discussion of the inverted spin. Key points here related to spin entry, the increased rate of spin rotation, and spin recovery. Two methods of spin recovery were discussed, the Muller/Beggs technique and the active spin recovery techniques. It is quite possible to pick the point at which you want to stop an inverted spin and hit it (or very close to it) using active spin recovery techniques. Pictures showing the views to the front and left during an inverted spin were used to give you an idea of what the plane is doing from a pitch standpoint during the spin. The chapter finished with a review of common errors pilots make during inverted spins.

Inverted spins are one of the most entertaining spin maneuvers to execute. Before you attempt inverted spins, get flight instruction from a qualified instructor, always use a properly certified aircraft, and wear a parachute. Anytime a plane is inverted, you are performing an aerobatic maneuver and you must use a parachute. Check the plane's weight and balance, give it a very thorough preflight, and then go have a great time!

7

Flat Spins

MANY A STORY HAS BEEN TOLD ABOUT THE FLAT SPIN. MOVIES GLO-
rify the hero pilot in a flat spin, fighting to get the plane out of it, at the last
moment lowering the flaps or landing gear to force the nose of the plane down.
Test pilot General Chuck Yeager's autobiography relates an incident in which
the F-104 Starfighter he was flying ended up in a flat spin, and only the ejection
seat and his experience saved him. It is no wonder that pilots view flat spins
with a certain amount of mysticism and awe. Airshow pilots thrill crowds each
year by executing the dangerous flat spin, with the crowd watching as the plane
spins like a frisbee. Even though few pilots have ever experienced one in person, most pilots have heard of them or have a story to tell about an anonymous
pilot that got his Luscome or Piper into a flat spin, never to recover.

There are good reasons for wanting to avoid flat spins. Many aircraft are
unrecoverable once they enter one. What actually causes a flat spin? How can
pilots reduce their chances of entering them? Are there airplanes that can execute flat spins and is there a way to recover from a flat spin? These are some of
the topics that this chapter covers. The theme that runs through this book continues here: stall/spin awareness, recognition, and avoidance. By flying a plane
correctly, you reduce the possibility of entering an unintentional spin, including
a flat spin.

Both upright and inverted flat spins are explained in this chapter. Pictures
are included to show forward and left-looking views of both of these spin scenarios. In the correct aircraft, with the proper instruction, it is quite possible to
execute flat spins safely. However, never, never attempt these in any airplane
except one certified to execute flat spins safely, and then only perform the spin
with a qualified flight instructor. No spin should be taken lightly; but under the
wrong conditions, flat spins can be disorienting and dangerous.

On a lighter note, why do airshow pilots like to execute inverted flat spins
as opposed to upright flat spins? The answer is that when smoke is used in the
airshow, upright flat spins result in a lot of smoke in the cockpit. The plane is

spinning like a frisbee, moving down into the smoke coming out of the exhaust system. Inverted flat spins put the smoke above the plane during the spin and reduce the amount of smoke that makes it to the cockpit.

DEFINING FLAT SPINS

In a number of previous discussions, the term "flat mode" was used in relation to various spins. In Chapter 4 the relationship between the rate of rotation and the pitch angle was discussed. As the nose of the plane rises, the spin's rate of rotation slows. This is referred to as the spin entering a flat mode. This flat mode can be the result of a number of factors, including aerodynamics, control inputs, engine power settings, and center of gravity.

Many planes certified for spins oscillate in pitch during the course of a spin, and in most cases this is not a dangerous situation. However, under the wrong circumstances a plane's nose does not oscillate down again after it rises. When this begins to happen, the plane is entering a flat mode. The change in the aircraft's attitude, a slowing of the spin's rate of rotation, and a reduction in the sound of air flowing by the plane are all indications that a spin is becoming flatter. Figure 7-1 illustrates a plane in both a normal, pitch down attitude during a spin and in a flat mode. As you can see, the nose of the plane is almost level in the flat mode. While I have never experienced it, some aircraft may have their nose rise above the horizon when in a developed flat spin.

A flat spin results when a plane enters a severe flat mode during a normal spin. The nose rises to pitch angles well above what they are through the course of a normal spin. The wing to the inside of the spin also rises, in some cases until the plane's wings are level in relation to the horizon. The rate of rotation slows, and the plane resembles a frisbee as it continues in the spin. The pilot now feels as though she is sitting in the middle of the frisbee, the airplane rotating flatly around her. From inside the plane, the pilot's view also changes, as compared to a normal spin. It becomes difficult to see the ground, because the nose is so high in comparison to a normal spin. This lack of ground reference points often causes a pilot to lose track of the plane's orientation and of how many turns the plane has executed during the spin. When doing flat spins, I use the best reference point available to keep track of the number of turns, which is the sun. Each time I see the sun pass the nose of the plane, I increase the spin turn count by one.

Some airplanes have a natural tendency to enter flat spins due to their design. These planes are typically placarded against all spins because once they have entered a spin, they may quickly enter a flat spin from which they may not be able to recover. This is one reason why it is so important to never spin a plane that is not certified for spins. You have no idea how the plane will react during a spin.

Chapter 3 contained the FARs pertaining to spin testing of normal category aircraft. You may remember the statement "A single-engine, normal category airplane must be able to recover from a one-turn spin or a 3 second spin, whichever takes longer, in not more than one additional turn, with the controls used in the manner normally used for recovery." During this brief exposure to

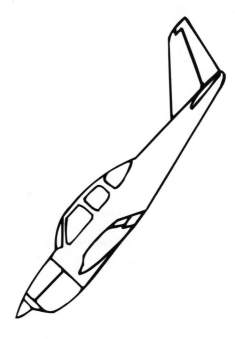

AIRCRAFT IN NORMAL MODE SPIN ATTITUDE

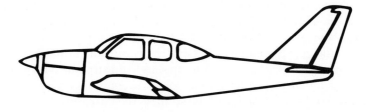

AIRCRAFT IN FLAT MODE SPIN ATTITUDE

Fig. 7-1 Normal versus flat spin modes.

a spin, the plane may show no poor spin characteristics until past the one turn or three seconds of spin. Past that point, the plane's nose may suddenly rise, causing the plane to enter a flat spin. In any case, never put a plane that is not capable of consistently recovering from a flat spin into one. Many aerobatic airplanes can safely execute flat spins, but check the operations manual or call the manufacturer to find out what the correct flat spin entry and recovery techniques are before you execute them.

What aerodynamic factors can cause a plane to enter a flat spin? Center of lift and pressure changes can cause a plane to enter a flat spin. Chapter 3 noted that having flaps down during spins may cause the spin to be flatter, due to the effect this has on the plane's airfoil. Many general aircraft operations manuals call for flap retraction immediately upon spin entry, which helps prevent the plane from entering a flat spin. Use of counter-spin ailerons by the pilot will also increase the chances of a spin flattening. Here is yet another reason why you should maintain neutral ailerons during a spin. Ailerons applied opposite the direction of an upright spin will cause the wing to the inside of the spin to rise, bringing the wings to level. This is one step in purposely entering a flat spin.

Use of engine power during the spin often causes the nose to rise, flattening the pitch attitude of the plane. The additional thrust provided by the engine will cause the nose to rise for many single-engine, general aviation aircraft due to the placement of the thrust line in relation to the center of gravity. For this reason, unless the plane's manufacturer notes otherwise, the first important step to spin recovery should always be to reduce the engine's power to idle.

Pilots who have accidentally entered an uncoordinated, power on stall may never think to pull the power to idle as they fight to avoid entering an unplanned spin. After all, they have been taught to increase power to full throttle to recover from stalls during flight training, so leaving it in when trying to recover from a spin must also be a good idea, right? WRONG! Leaving the power at full throttle may be just the ingredient that causes the plane to enter a flat spin from which it may never recover. Chapter 2 discussed how adding engine power during stall recovery will raise the nose; this axiom also holds true for spins.

Aft center of gravity can also cause a plane to enter a flat spin. Figure 7-2a shows a plane with an aft center of gravity. As the plane begins to rotate during the spin, centrifugal force will attempt to force the center of gravity to the outside of the circle created by the plane. The following example will help you visualize this. Imagine tying a bucket to one end of a rope, then twirling the bucket around your head by swinging the rope from the other end. The faster you swing the bucket, the more it will move to the outside of the circle created by the swinging rope. This same principle applies to an airplane with an aft center of gravity. As the plane begins to rotate, the aft center of gravity and the tail are forced down and to the outside of the spin. As the tail drops, the nose of the plane pitches up, causing the spin to flatten. See Figure 7-2b, in which the plane is almost level as it spins in a flat mode.

Chapter 8 covers NASA studies and NTSB accident reports that illustrate, in a deadly manner, how important it is that you remain within a plane's

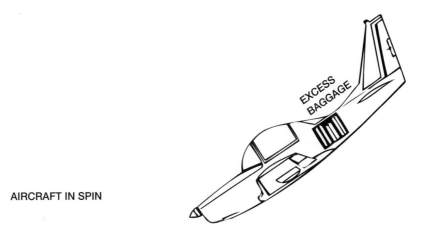

AIRCRAFT IN SPIN

Fig. 7-2a Aft center of gravity.

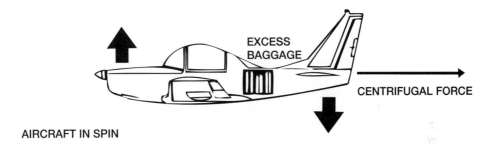

EXCESS BAGGAGE

CENTRIFUGAL FORCE

AIRCRAFT IN SPIN

Fig. 7-2b Centrifugal force creates flat spin mode.

weight and balance whenever you do any spin training. More than one spin accident has been caused by the airplane being out of the center-of-gravity (CG) envelope. In these situations, even a plane that is very docile during a spin may become unrecoverable.

Chapter 6 discussed inverted spins. It is quite possible to enter inverted flat spins as well as upright flat spins. In fact, while giving aerobatic instruction, I have had aerobatic students put the plane into accidental inverted flat spins much more frequently than they have put the plane into accidental upright flat spins. In either case, proper spin recovery techniques allow for rapid recovery. As you read this chapter, keep that point in mind. When done in the correct airplane, with the proper instruction, inverted flat spins are no less safe than upright flat spins and I regularly teach aerobatic students how to execute both types of flat spins.

The biggest challenge facing most students during advanced spin training is learning to recognize that they are in a spin when the plane has entered a flat spin. Due to the flat attitude of the plane and the decidedly different motions the plane is executing as compared to a normal spin, many pilots do not realize they are in a spin at all, let alone a flat spin. This is a real concern. If you do not recognize the fact that you are in a spin, trying to fly the plane out of the situation will prove very difficult. Any control inputs you might make that are not the correct recovery procedures may only make the situation worse. Proper spin recovery actions must be taken to recover from a flat spin.

Like inverted spins, I like doing flat spins. They are fun to execute and give pilots exposure to another set of spin-related unusual attitudes. Knowing how to enter and recover from them, how to recognize them, and how to avoid them increases a pilot's ability to fly safely. The remainder of the chapter covers flat spin entry and recovery for both upright and inverted flat spins. The pictures included of these spin situations should give you an idea of what it looks like from inside the plane during these spins. As you look at them, pay close attention to the attitude of the plane. If you did not know these still pictures were taken from a plane in a flat spin, you probably would think the plane was in level flight. This is one indication of how different flat spins look from normal spins, and what makes them more difficult for pilots to recognize them.

UPRIGHT FLAT SPINS

Upright flat spins are really an awful lot of fun when done in the right airplane, with the proper flight instructor. Most spin students are able to successfully enter and recover from planned upright spins after only a few attempts. The most common problem pilots have when learning flat spins is the proper use of controls. As you will see, they are positioned in an unusual manner that you must get used to. Once you understand the correct control input sequence, flat spin entry becomes easy. Flat spin recovery relies on the well-documented emergency spin recovery procedure, the Muller/Beggs technique. This has proven to be extremely effective in the Pitts S-2B and some other aircraft. For that reason, I teach this recovery method to students learning flat spins.

Chapter 3 introduces normal, upright spin entry, which was then reviewed in Chapters 4 and 5 as part of the spin discussions. Normal spin entry techniques are used to initiate upright flat spins as well. When practicing flat spins, always give yourself as much altitude as possible. It is not uncommon for me to have students climb to 9,000 feet AGL for flat spin practice.

There are several reasons for this additional altitude. First, you can never have too much sky underneath you when you are learning a new maneuver. Given some of the problems pilots can experience, having as much room between you and the ground as possible is wise. Second, flat spins represent a completely new spin mode for pilots. When they are first learning to execute a flat spin, pilots are concentrating on many new aspects related to this spin.

Control inputs, throttle settings, the new view outside the aircraft, motions the plane is going through, and how the flat spin is making them feel must all be experienced. While I always stress altitude awareness during spins, pilots occasionally become so preoccupied with other factors that they forget about keeping a watch on the altitude. The third reason to practice flat spins at high altitude is that depending on the plane, the weight, and balance, recovery from flat spins can take a fair amount of altitude. You do not want to come up short.

Once you have reached a safe altitude, do your normal clearing turns, set carburetor heat as required, and then pick a cardinal heading to fly as you slow the plane for spin entry. As always, reduce power to idle, and then maintain altitude by easing back on the elevator control while you bleed airspeed off. When the plane stalls, input full rudder in the direction you want to spin and ease the elevator control fully aft. At this point in the spin, be sure to keep the ailerons centered in the neutral position.

When students are first learning flat spins, I have them complete two turns before beginning flat spin control inputs. This allows the plane to settle down into a developed spin. During the incipient spin phase, control movements can sometimes prevent the plane from entering a developed spin and make flat spin entry more difficult. Once the plane has completed two turns, control inputs are made in the following order:

1. While keeping the elevator control in the full aft position, input full counter-spin aileron. For example, if you are executing a flat spin to the left, you input full right aileron. Maintain full pro-spin rudder input during the flat spin.
2. Increase the throttle to bring in engine power. How much power is necessary depends on a number of factors, such as engine power, aircraft weight, and center of gravity. I have students use enough to bring the nose of the plane level with the horizon. Some planes may not be able to bring the nose level with the horizon, even with full engine power.
3. Maintain the controls in this position and enjoy the ride.

As noted earlier, counter-spin aileron input will raise the wing to the inside of the spin. As you make this input, notice just how much this aileron input can affect the plane. Most pilots are surprised at how much this seemingly minor control change affects the attitude of the plane while in the spin. Contrary to what some pilots think, adding counter-spin aileron will not get them out of the spin. It instead serves to flatten the spin the plane is already in. You may also notice that the nose of the plane may tend to rise as a result of the aileron input. How much aileron you should use to level the wings varies based on the plane you are flying. In the Pitts S-2B, I use full counter-spin aileron and this works very well in setting up the flat spin. Other planes may not require full aileron input to bring the wings to level; while some, even with full aileron input, may not be able to level the wings.

As you add engine power, the nose of the plane will rise. How much it rises depends on engine horsepower, the center of gravity, and other factors. As the nose of the plane rises, it may also cause the wing to the inside of the spin to rise as well. I have heard some pilots say that they can get the nose of the plane well above the horizon as power is added. In the Pitts Special, I typically use about one-third to one-half throttle to get the nose level with the horizon, depending on how the plane is loaded. Pilots first learning flat spins have expressed their shock in no uncertain terms at what adding power during the spin does to the aircraft's pitch attitude.

Once the control and power inputs have been made, the plane will continue to rotate in a flat spin. The rate of rotation will typically be slower than in a standard spin, and the sound of air flowing around the plane will also probably be quieter. While in the spin, remember to use some reference point such as the sun to keep track of how many turns you have completed.

Also, be sure to keep track of your altitude. I typically have students begin recovery at 6,000 feet AGL at a minimum. During the first exposures to flat spins, we usually execute three or four turns in the flat spin, and then recover. For most pilots, this is more than enough, as they are overwhelmed by the actions of the plane during the spin. While you are in the flat spin, look around. During their first flat spin lessons, most pilots tend to focus straight ahead of the plane during the spin, limiting their field of view. This reduces their ability to grasp as many aspects as possible related to flat spins. The more you take in, the better able you will be to recognize flat spins, and recognition is the key to recovery.

The Muller/Beggs emergency spin recovery technique is the preferred method I use and teach for flat spin recovery. Just as with accelerated, crossover, and developed inverted spins, it may be very difficult to position the controls correctly for recovery. At an aerobatic safety seminar I attended, flat spin recovery was one of the topics of discussion. The speaker indicated he had used active spin recovery to recover from flat spins, until one day he could not find the right positions for the controls. Finally able to recover from the spin, he decided to take a more serious look at the Muller/Beggs technique. To review, the recovery steps are:

1. Cut that throttle!
2. Take your hands off the stick!
3. Kick full rudder opposite until the spin stops!
4. Neutralize rudder and pull out of the dive!

Cutting the throttle begins to drop the nose of the plane, releasing the control stick allows the airstream to position the controls in the neutral position, and using full rudder opposite the spin direction stops the yawing motion of the spin. The plane will end up in a dive, just as during recovery from other spins. Remember, though, that the Muller/Beggs technique may not work for all planes. If the manufacturer of the plane you fly requires a different recovery technique, use that. Whatever recovery technique the plane you fly requires, once you are out of the spin, fly the plane back to level flight and put a big grin on your face. Figures 7-3a through 7-3e and Figures 7-4a through 7-4c show the forward and side views during the life of a flat spin. Note the pitch attitude changes as counter-spin aileron is applied. During recovery, the plane comes out of the flat mode, allowing the nose to drop. At that point, the plane is flown back to level flight.

Fig. 7-3a Flat spin entry: forward view.

Fig. 7-3b Flat spin normal mode: forward view.

Fig. 7-3c Flat spin flat mode: forward view.

Fig. 7-3d Flat spin recovery: forward view.

Fig. 7-3e Flat spin straight and level: forward view.

Fig. 7-4a Flat spin entry: left side view.

Fig. 7-4b Flat spin normal mode: left side view.

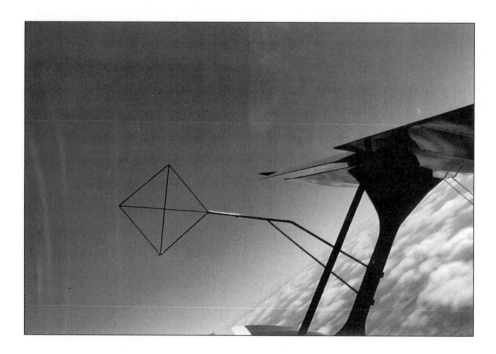

Fig. 7-4c Flat spin flat mode: left side view.

Fig. 7-4d Flat spin recovery: left side view.

Fig. 7-4e Flat spin straight and level: left side view.

Common Errors

Pilots learning flat spins can make several errors during both entry and recovery. Related to spin entry, incorrect application of aileron can slow or prevent the plane from entering a flat spin. Due to confusion over proper aileron input, pilots will occasionally apply pro-spin ailerons, putting the plane in a knife-edge spin. They may also fail to input any counter-spin aileron, or less aileron than necessary to bring the wings to level or thereabouts. In addition to applying aileron, pilots must also remember to maintain the control stick in the fully aft position as ailerons are applied.

With the unusual use of controls, pilots sometimes let the control stick move forward as they input aileron, allowing the spin to accelerate. They may also reduce the amount of pro-spin rudder input. As far as throttle settings are concerned, pilots may be tentative about applying throttle, or may use too much throttle. The amount varies depending on the plane, so consult the manufacturer if you have any questions related to throttle use.

All of the errors associated with spin recovery can occur with flat spins. The big item to keep in mind is to COMPLETELY retard the throttle to idle when recovering. Pilots occasionally pull the power back, but not to idle. Even small amounts of power can slow or prevent your ability to recover from the flat spin. Also be certain to use full counter-spin rudder input when you recover. Remember to give yourself lots of altitude to recover; don't cut things close when it comes to safety.

INVERTED FLAT SPINS

Inverted flat spins are just as much fun to execute as upright flat spins. I have found that aerobatic students have a tendency to put the plane into accidental inverted flat spins while practicing certain maneuvers. I also think the spins look pretty cool from inside the plane, but that could be due to my predisposition for inverted flight. This section reviews how to execute inverted flat spins. You will find many similarities between inverted and upright flat spins, with the minor differences due to the fact that you are inverted.

First, what is an inverted flat spin? It is the same maneuver as the upright flat spin, except the plane is inverted. The plane has entered a flat mode, inverted spin and still resembles a frisbee as it spins flatly around. The wings are parallel, or nearly so, to the horizon. The nose has risen from its normal inverted spin pitch attitude to one that is level with the horizon. From inside the plane, pilots feel like the plane is spinning flatly around them. The physical feelings a pilot experiences in both upright and inverted flat spins are decidedly different from those experienced during normal spins. The g-forces seem less than in a normal spin, and the plane's rate of rotation is less than that of a normal inverted spin.

As with inverted spins, the inverted flat spin is initiated from an inverted stall. Before beginning the half roll to inverted, execute a series of clearing turns, and then do a half roll to inverted. Apply carburetor heat as necessary, and then reduce power to idle. Hold altitude by easing forward on the control stick, using rudder as necessary to maintain coordinated flight. As the plane stalls, apply full rudder in the direction you want the plane to spin, easing in full forward elevator input as you do so.

I typically have students enter the inverted flat spin after the plane completes about two turns of the inverted spin. As you will recall, this allows the plane to get through the incipient stage and into a developed spin. The steps to put the plane into the inverted flat spin are:

1. While keeping the elevator control in the full forward position, input full pro-spin aileron. For example, if you are executing a flat spin to the left, you input full left aileron. Maintain full pro-spin rudder inputs during the flat spin.

2. Increase throttle to bring in engine power. How much you add depends on a number of factors, such as engine power, aircraft weight, and center of gravity. I have students use enough to bring the nose of the plane level with the horizon.

3. Maintain the controls in this position and enjoy the ride.

Apart from being inverted, the plane behaves in much the same manner during the inverted flat spin as it does in the upright flat spin. The nose will be nearly level with the horizon, as are the wings. The rate of rotation is typically slower than during an inverted spin. Be sure to look around outside the plane during the spin to absorb what the plane is doing. The top of the cowling (although during the inverted flat spin, it is on the bottom of the plane) on my Pitts Specials have a tendency to "oil can" in as a result of air pressures during the inverted flat spin. Figures 7-5a through 7-5c show the forward view from the plane during the stall/spin/flat spin progression. Figures 7-6a through 7-6c show this same sequence while looking to the left of the plane.

Spin recovery once again is performed using the Muller/Beggs technique. Reduce the throttle to idle, release the stick, and input full rudder opposite the direction of the spin. The plane will recover in an inverted dive, from which you fly back to straight and level flight. Be sure to avoid excessive altitude loss when recovering from the dive. Do not let the airspeed build too rapidly and do not pull too hard on the elevator controls, which would set up the potential for an accelerated stall to take place as you make your way back to level flight. Figures 7-7 and 7-8 show the plane as it recovers from the spin, both with forward views and views to the left.

During the spin, pilots can have many of the same disorientation problems that they have during upright flat spins. It can be difficult to pick out the horizon or reference points on the ground. Keeping track of the number of turns is therefore more difficult. Add to this the negative g's, and inverted flat spins represent yet another new flight experience. The lack of reference points, and difficulty with seeing the horizon, is why it is so important to look around during the spin. This will help you find the correct places to look during the spin to determine the plane's attitude. Using the sun to help you keep track of the number of turns can also be useful during inverted flat spins.

Fig. 7-5a Inverted flat spin entry: forward view.

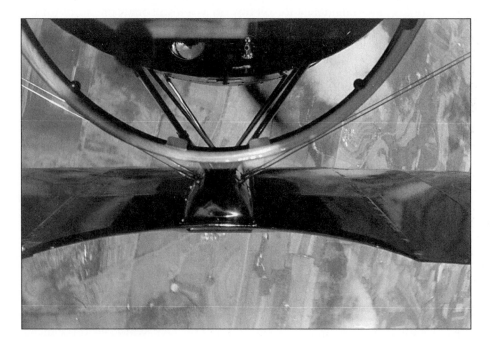

Fig. 7-5b Inverted flat spin normal mode: forward view.

Fig. 7-5c Inverted flat spin flat mode: forward view.

Fig. 7-6a Inverted flat spin entry: left side view.

Fig. 7-6b Inverted flat spin normal mode: left side view.

Fig. 7-6c Inverted flat spin flat mode: left side view.

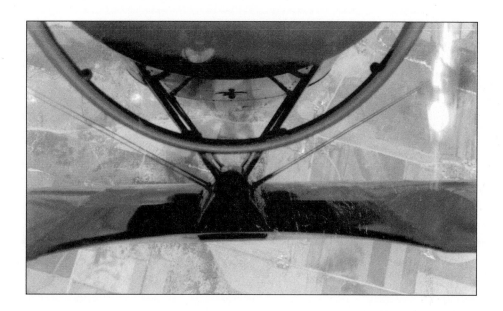

Fig. 7-7a Inverted flat spin recovery: forward view.

Fig. 7-7b Inverted flat spin straight and level: forward view.

Fig. 7-8a Inverted flat spin recovery: left side view.

Fig. 7-8b Inverted flat spin straight and level: left side view.

188   *Flat Spins*

Common Errors

Most of the errors seen with upright flat spins apply to inverted flat spins. During entry, pilots will sometimes let the stick move back, potentially recovering the plane from the inverted spin or putting it into an accelerated spin. Given that for upright flat spins, aileron input is made opposite the rudder input, pilots occasionally use this same control configuration in an attempt to enter inverted flat spins. This puts the plane into an inverted knife-edge spin to some degree, depending on the aileron authority of the plane, weight and balance, and other aerodynamic factors. At any rate, make the aileron input in the same direction as your rudder input. You do this for the same reason that you use opposite aileron and rudder when making inverted turns. Chapter 6 covered a situation in which a pilot used rudder opposite the aileron input to make a coordinated control input. To get the cross-control between the ailerons and rudder necessary for an inverted flat spin, you must use aileron in the same direction as rudder.

Some airplanes seem to enter inverted flat spins more readily when power is added before you add pro-spin aileron. Pilots have mentioned that this helps the plane they fly enter the flat spin more smoothly, so you may need to try this alternate technique to see how well it works.

During recovery, make sure you get the throttle completely to idle. I stress this because if you forget to pull the power to idle, or only get the throttle partially back, the plane may not recover from the flat spin, even with the correct control inputs. Make sure you give yourself plenty of altitude both during entry and recovery. Initially, use altitudes at or above those you use for upright flat spins. Make a point of looking outside the plane while you execute the inverted flat spin, looking for as many external reference points as possible. Due to the unusual nature of inverted flat spins, and how difficult it can be to keep track of what the plane is doing, keeping your head inside the cockpit can make it difficult to "re-acquire" the horizon or ground references when you do look back outside again.

Altitude awareness is also extremely important. I have read accounts of pilots that are so enthralled by the experience of a flat spin that they forget to monitor their altitude and are unable to recover due to a lack of altitude. Once again, because this is an aerobatic maneuver, be sure you are wearing a parachute that is properly packed. In fact, once you move beyond normal spin training, you should wear a parachute for any spins you attempt.

SUMMARY

Flat spins are a maneuver that many pilots have only heard stories about. Flat spins are often considered deadly and unrecoverable in any situation. In the right aircraft, this is not true. Flat spins are fun and add yet another dimension to unusual attitude recovery and spin recovery. While you may not be able to get flat spin training at your local airport, understanding what factors can cause a plane to enter flat spins can help you avoid them.

Center of gravity, engine power, and aileron inputs are all factors that you can control when you fly an airplane. Knowing how to compute the weight and balance for the plane you fly, and ensuring that the plane is within its weight and balance limits can help prevent a flat spin from developing while you are in a normal spin. Reducing engine power to idle immediately when faced with a spin will also help prevent the spin from entering a flat mode.

You must maintain neutral ailerons during a spin. Aileron input can cause a spin to flatten if counter-spin ailerons are used, or to enter a knife-edge spin if pro-spin ailerons are input. A safety inspector with the FAA related an accident investigation he was involved in to me. The single-engine plane involved in the accident had impacted the ground in a flat, spinning attitude. The investigation showed that the pilot had been attempting to stop the turning using counter-spin aileron, causing the plane to enter a flat spin. Recovery from the spin was never completed and it appeared the pilot maintained counter-spin aileron until impact with the ground.

If you decide you want to pursue flat spin training, be certain it is in a plane that is capable of safely entering and recovering from flat spins. This restriction immediately narrows the list of usable aircraft down to only the most advanced aerobatic planes. I give upright and inverted flat spin instruction in a Pitts Special S-2B, but there are other similar aircraft that can be used. If a flight school attempts to convince you that their normal aircraft can be used for this type of training, head in the other direction. Use parachutes during flat spin training and be certain you have been given a thorough briefing by your instructor related to control inputs for the plane you are flying, both for spin entry and recovery. The control inputs for spin entry and recovery covered in this chapter may work with the plane you fly, but if the manufacturer recommends different procedures, use those. Your instructor should also brief you regarding emergency exit procedures, and use of the parachute, in the event the aircraft suffers damage or becomes uncontrollable.

The remainder of the book is dedicated to review of accidents related to stall and spin situations, spin studies that NASA has completed on general aviation aircraft, and tips to instructors on how to give spin training.

8

Spin Studies and Stall/Spin Accidents

THIS CHAPTER IS A DEPARTURE FROM THE "HANDS-ON" APPROACH taken in the book to this point. You need to begin to understand how serious stalls and spins are under the wrong circumstances, and how important it is to be "situationally aware" while you are flying a plane. Allowing the plane to enter a stall or spin at low altitude can put you in an unrecoverable situation, no matter how proficient you are at recovery. The first portion of the chapter reviews a number of different stall/spin accidents on record with the NTSB. As you read through them, you will see that all of the accidents were avoidable, if only the pilot had been more aware of what the plane was capable of, or not capable of, depending on the circumstances.

The second portion of the chapter reviews spin studies that NASA has done on general aviation aircraft. This section takes a more detailed look at what NASA studies have found. Once again, understanding what causes a plane to stall or spin can help you avoid getting into them under normal circumstances. While most people associate NASA with the manned spaced program, NASA has been involved in aerodynamics related to general aviation for many years.

ACCIDENT 1

This accident is reviewed first for two reasons. First, it involves an aerobatic airplane. Second, the plane was loaded outside of its center-of-gravity envelope. As a result, a plane that is normally quite capable of dealing with spin situations was rendered uncontrollable during a stall/spin. In this incident, a Christen Eagle II lost control during acrobatic maneuvers, collided with coastal waters off the coast of California, and was never recovered. Witnesses observed the plane in a steep climb, followed by a wingover, and then a spin. Witnesses further stated that the pilot attempted to pull out of the spin after four to five rotations. Instead of recovering, the plane continued in a spin, simultaneously flipping nose over until its tail struck the water.

The pilot had fourteen flight hours of experience in the aircraft make and model. Accident investigators found that the airplane's weight was in excess of both the normal and aerobatic allowable gross weights. They also determined that the center of gravity exceeded the aft limit of the Eagle's aerobatic allowable CG. The accident report further states the airplane's flight manual contains a warning that states: "Pilots must take weight and balance responsibilities seriously. All aircraft must be loaded so that the aircraft CG will remain within safe limits at all times. All spin recovery procedures depend on having the CG remain within design limits."

This example should clearly point out the importance of verifying you are within acceptable center-of-gravity ranges when you are executing spins. The Christen Eagle is a fine aerobatic plane. But in a situation where it was being flown outside the center-of-gravity envelope, it became unrecoverable during a spin. So, even if you think you know your plane, take the time and compute the weight and balance. Besides brushing up on how to do weight and balance calculations, you may be surprised by how sensitive some aircraft are to weight and balance. Some aircraft have very narrow center-of-gravity ranges and can easily be loaded outside those ranges.

Six more accident scenarios follow, each an example of situations discussed in the book. As you read each of them, think about how you might avoid that situation.

ACCIDENT 2

In this accident, the pilot was a student flying solo in an experimental, home-built aircraft. The pilot climbed to an altitude of 4,200 feet MSL, intending to practice spin entry and recovery. Out of curiosity, the pilot applied engine power while in the spin to see if the plane would spin faster. The airplane's nose immediately rose, and the plane entered a flat spin. At that point, the plane's engine lost power and the pilot was unable to recover prior to the airplane impacting the ground.

Chapter 7 covered this very situation. Adding power when in a spin usually causes the nose to rise, potentially putting the plane into a flat spin that it will not be able to recover from. In this accident situation, the pilot clearly did not understand what would happen. The information contained in the accident investigation did not indicate what spin recovery techniques the pilot tried. The plane may not have been recoverable once in a flat spin; but again, the accident report does not indicate if the plane, a Cleveland Piel Emeraude, is able to successfully execute recovery from flat spins. If the pilot had been aware of the affect of adding engine power during a spin, this situation could have been avoided.

ACCIDENT 3

This example shows what can happen when a pilot loses track of what the plane is doing, ignoring airspeed and the proper use of controls. In this particular incident, the pilot was scouting a hunting area with passengers, attempting to find likely hunting spots. During this scouting flight, the pilot was maneuvering the plane at low altitudes in and around a canyon, while observing the terrain. During this time, the pilot's attention was diverted from flying the plane, and the airspeed began to deteriorate. The plane stalled, then spun into the ground.

Several different chapters of this book have discussed how important it is to always maintain control of the plane. This is especially true when you are flying close to the ground at slow airspeeds, such as during takeoff or landings. This incident is a classic example of a pilot being distracted from flying the plane by his passengers or the view outside the plane. If the pilot had been more aware of sounds the engine was making, the decreasing sound of the air flowing over the plane, or the increasing softness of the controls as airspeed bled off, he might have noticed the reduction in airspeed long before the plane stalled. Never divert your attention from flying the plane so much that you do not pay attention to these sensory inputs.

One way to become more proficient at "feeling" the airplane is to practice flying at minimum controllable airspeeds to become accustomed to how the airplane behaves in this flight regime. Doing stalls or slow flight every two years during the biennial flight review hardly allows a pilot to maintain proficiency at these maneuvers. Like any skill, you must practice it often to become better at it, and flight at MCA, including stall awareness, recognition, and avoidance, is no exception.

ACCIDENT 4

This flat spin accident was caused by a different set of circumstances. The plane, a Piper PA-38-112, was being flown by a flight instructor and primary student. They were flying at minimum airspeeds at an altitude of approximately 650 feet AGL. The airplane stalled with the flaps fully extended, entered a flat spin, and impacted the ground.

While this accident report summary was not long, it contained information that reflects yet another situation covered in this book. If you enter a spin with flaps extended, the plane tends to enter a flat spin. Many aircraft operations manuals indicate that if the plane enters a spin, the flaps should be retracted to avoid this situation. Just as in accident 2, in which the pilot wondered what adding power would do while the plane was in a spin, many pilots do not realize what effect having flaps extended will have on a plane during a spin.

A second point to consider in this example is why were the flight instructor and student practicing flying at slow airspeeds only 650 feet AGL? Given that the plane stalled and entered a spin, this may have been a salvageable situation if the pilots had been at a higher altitude. Even a one-turn spin typically uses 1,000 feet of altitude from entry to recovery under the best of circumstances. If you are going to practice slow flight, do it at an altitude that allows you to recover from mistakes.

One might also wonder whether the pilots were flying the plane in a coordinated manner when it stalled. The fact that it entered a spin indicates there was some yawing input at the time of the stall, causing the plane to spin. The four left-turning tendencies (P-factor, torque, slipstream, and precession) must all be accounted for when flying a plane at MCA. Keeping the ball centered may also have prevented this accident. If a spin had been avoided and the pilots had needed only to recover from a stall, they may have had sufficient altitude to safely recover.

ACCIDENT 5

In this accident, the pilot was flying a glider, setting up for landing. The plane experienced higher-than-normal sink rates on downwind, ending up at a lower-than-normal altitude when abeam the end of the runway. Rather than compensate for this unusual loss of altitude by turning to base early, the pilot elected to continue on downwind and turned to the base leg of the pattern at the location relative to the runway at which he normally turned base. Once on base, the pilot realized he would have problems reaching the runway and attempted to compensate by tightening the turn from base to final in hopes of making the runway. During this tight turn, the pilot cross-controlled the glider, which stalled and then spun into the ground.

The next chapter emphasizes that most accident situations result from a series of events that put the pilot in continually worsening situations. In this accident, you can identify a number of events that led up to the stall/spin. If the correct decision had been made at the proper time, the pilot had a number of opportunities to avoid this particular spin. First, if he had recognized and, more importantly, reacted to the increased sink rate while on downwind, the issue of coming up short of the runway could have been avoided. Instead, the pilot used location from the runway, as opposed to proper touchdown zone sighting techniques, to mark where the turn to base should take place. A turn to base that compensated for the lower altitude would have allowed him to make the runway.

Second, once on base he attempted to compensate for the altitude loss by tightening the turn to final. Earlier chapters discussed how dangerous tight turns at low altitude and slow airspeeds are. Stall speed increases as a result of increased bank angles, in some bank situations rising well above the normal stall speed of the airplane. Finally, the pilot cross-controlled the airplane, setting it up for a spin when the plane stalled. This shows how important it is to maintain coordinated use of the controls to avoid spins. In addition, planes that are in slips or skids are going to lose altitude faster as a result of the additional drag the cross-controlled situation sets up.

If you are short on altitude, keeping your turns coordinated will help reduce your altitude loss. Planes can usually recover from a stall with much less altitude loss than they can from a spin. In this particular accident, the pilot may not have been able to make the runway, but he could have had other potentially safe landing areas within gliding distance. Avoid locking your attention on one item, such as the runway, thereby reducing the number of options you have available due to a lack of awareness.

You may be thinking this situation could never happen to you, but the fact that fairly large numbers of pilots are getting themselves into stall/spin accidents every year means it's happening to someone. Be aware of what your plane is doing, keep flying it at all times, and keep thinking, especially when the situation gets beyond normal flight parameters.

ACCIDENT 6

This accident is included because I have flown with a number of pilots who indicate they like to rudder the nose around during turns, as opposed to banking the airplane. The previous accident example highlights what cross-controlling can do to a plane. This accident adds a new twist to the situation.

In this accident, the pilot was flying a Cessna 152 equipped with an FAA-approved short takeoff or landing wing modification (STOL). Shortly after takeoff, the airplane was observed by witnesses turning as if to return to the runway for a landing. The plane was observed with the wings level, but moving sideways, the flaps fully extended. The plane stalled and then spun into the ground. Investigators found that the pilot had a habit of making wing-level turns using rudder, which is clearly an uncoordinated turn situation. During the investigation, they also found that the wing flow fences, part of the STOL modification, were incorrectly installed. Evidently, there was a partial magneto failure that resulted in a partial loss of engine power.

A series of events can be identified that resulted in this accident. First, it appears the engine suffered partial power loss as a result of a magneto failure. The report does not indicate if the pilot had done a run-up prior to takeoff. Some pilots are in the habit of assuming that the magnetos are working fine and do not bother to test them prior to takeoff. The confidence pilots have is well founded because magnetos rarely fail. In twenty-plus years of flying, I have had only one magneto fail, but I found it during run-up.

The pilot then attempted to return to the runway for landing after the problem was discovered. Once again, the report does not tell us what the altitude during the turn was, or how much power the plane had available from the engine. But conventional wisdom has shown that pilots should not make a steep, tight turn at slow airspeeds in an attempt to get the plane back to the runway. Very often there is a clear field or other area suitable for landing ahead of the airplane that would make a much safer landing possible. Do not narrow your field of vision too much by concentrating on one thing and ignoring other options.

The report does not confirm it, but the plane might have been able to make it back to the runway. The pilot was attempting to turn the plane using rudder, which increased the chances for a stall/spin. Until pilots experience a stall/spin from a skidding turn, they do not realize how quickly this can take place. I have taken pilots up to a safe altitude and had them execute skidding turns to the point at which the plane stalls. Invariably it stalls well before they expect it to; and when the stall happens, it breaks very rapidly, almost always entering a spin before they are able to recover. These pilots are briefed on what to expect, and yet they are surprised by the sudden onset of the spin. Imagine being at a low altitude, having your attention focused on concerns like the engine failure or making it back to the runway, and then entering a skidding stall. The odds for a successful recovery are not in your favor. The best way to avoid this situation is to fly the airplane in a coordinated manner; and when an emergency situation arises, keep all of your options in mind as you take action.

ACCIDENT 7

This example was included to show the importance of having a thorough understanding of spin recovery when doing aerobatics. Aerobatic students often accidentally put the plane into flat spins during aerobatic practice. In this accident, the pilot was practicing the Immelman maneuver. An Immelman is a half loop, followed by a half roll. When completed, the airplane is traveling in a direction opposite to what it was during entry to the maneuver. The half roll at the top of the loop is done at a slow airspeed and requires a certain amount of finesse on the ailerons and rudder. In fact, the maneuver is nicknamed the "Immelspin," due the fact that pilots new to the maneuver frequently end up stalling and spinning as they execute the roll portion of the maneuver.

In this particular accident, the pilot did just this as he executed the Immelman. As a result, the plane entered a flat spin at an altitude of 3,500 feet and impacted the ground. The pilot told investigators he attempted to recover, but was unsuccessful. A spin from the Immelman maneuver can easily set up pilots for a flat spin due to the fact that the engine will be at a high power output and the controls may be uncoordinated during the roll. These are all necessary ingredients for a plane to enter a flat spin. If you find yourself in any power on spin situations, as soon as the plane enters a spin with power on, you should immediately reduce power to idle, which will help prevent the spin from turning into a flat spin. If you intend to fly aerobatics, be sure you get adequate spin training. You should become very comfortable with spin recovery procedures and you need to be able to recognize and recover from spins immediately.

ACCIDENT REVIEW SUMMARY

This chapter has examined several examples of stall/spin accidents that took place between January 1, 1993, and January 1, 1995. The report used for this chapter from the NTSB contained a total of ninety-six accidents. That is an incredible number of accidents related to spins during a two-year period. Several themes ran through the accident reports. Pilots used improper control techniques, failed to correct for situations early enough in a series of events to prevent an accident, made bad decisions when they did take action, or just simply quit flying the airplane and let events take control of the situation.

You—not the airplane, not the FAA, not your passengers—must always be in control of and react properly to a given situation. In many cases, serious accidents can be avoided through knowing how to fly your plane properly. This means knowing how it reacts during stalls and slow flight, how to properly set up for landings, and adapt to changing conditions. If you do not feel comfortable with any area of flying, you should receive instruction from a properly qualified instructor. Not only will you gain confidence in how to react to situations, but you will become a safer pilot.

NASA SPIN STUDIES

The National Aeronautics and Space Administration has invested time and effort in studying the spin characteristics of aircraft. These studies have included civilian, air transport, and military planes. The remainder of this chapter reviews a NASA study pertaining to general aviation aircraft spin characteristics. This study deals with spin characteristics that NASA personnel have compiled from a number of different studies completed several years ago. The report indicates that little spin research for general aviation planes seems to have been accomplished in recent years, perhaps because few modern aircraft are certified for spins and spin training is no longer required for private pilots.

This particular report, while quite good, was done in 1971. Perhaps the FAA should reconsider this stance, given the number of spin-related accidents and the general lack of knowledge most pilots seem to have about spins.

Compiled by James S. Bowman, Jr., the paper *Summary of Spin Technology As Related to Light General-Aviation Airplanes* contains a great deal of information related to the spin characteristics, and more importantly, the factors that affect an airplane's ability to recover from them. It states that "three factors are found to be of almost overriding importance in spinning for this type of airplane." The relative distribution of the mass between the wing and fuselage, the density of the airplane relative to that of the air, and the tail design all play a factor in how general aviation aircraft react during spins and spin recovery.

To begin, the report defines spins as involving "simultaneous rolling, yawing, and pitching motions while the airplane is at high angles of attack and sideslip." It goes on to state that the "spin is maintained by a balance between aerodynamic and inertia forces and moments." In other words, the plane stays in a spin because the factors that caused it to enter the spin have stabilized, with the resulting forces stabilizing the plane in the spin. In order to recover the plane from a spin, some action must be taken to change those aerodynamic and inertia forces and moments to stop the spinning motion.

Mass distribution between the wings and fuselage is very important, as this affects what the correct spin recovery control inputs should be. Coupled with the density of the plane relative to the air, this "determines the tail-design requirements for recovery." Mass distribution determines how the spinning airplane will respond to control inputs, "especially to elevators and ailerons." Figure 8-1, from the paper, illustrates how the different mass distributions can be classified. In the fuselage-heavy loading, most of the mass is distributed along the fuselage. As you can see, in this case the recommended spin recovery procedure is ailerons in the direction of the spin and rudder against the direction of the spin. Most general aviation aircraft fall into the zero loading category, in which mass is distributed throughout the structure of the plane. In zero loading, the recommended spin recovery is rudder against the spin, followed by elevators down. The third category, wing-heavy loading, indicates that the greatest mass is distributed along the wings. The location of engines and fuel tanks in multi-engine aircraft often puts them in this category. Here the recommended recovery procedure is elevators down, plus rudder against the direction of the spin.

The report indicates that the loading of an airplane will dictate the control movements necessary for spin recovery. "Deflection of the rudder to opposite the spinning rotation directly is always recommended, but in many cases, it is not adequate to provide recovery." With the wing-heavy loading, down elevator becomes the primary control used for recovery. When an aircraft is fuselage-heavy, the aileron becomes the primary recovery control, but in this case aileron is used with the spin. ". . . for example, stick right for a right spin." When an aircraft is zero loading, the most important control for recovery becomes the rudder.

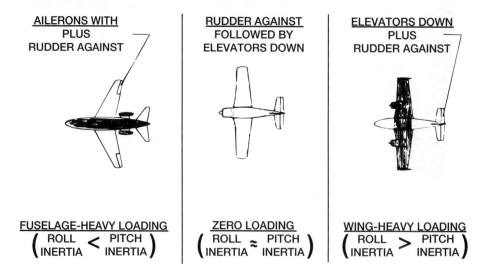

Fig. 8-1 Aircraft mass distributions.

The report indicates that aircraft that have a high relative-density factor require more rudder and elevator effectiveness, as compared to low relative density airplanes, to provide for successful spin recovery. In order for a plane to recover from a spin, relatively large moments are needed to overcome the inertia of a spin. This is even more true when planes are in flat spins and requires that the control surfaces be effective while the plane is in a spin. In order to accomplish this, a large amount of rudder must remain free of the horizontal tail wake. Figure 8-2, also from the study, illustrates how placement of the horizontal stabilizer near the middle or top of the fuselage necessitates that the rudder extend below it in order for it to remain effective during spins. When the horizontal stabilizer is near the bottom of the fuselage, the upper portion of the rudder must be designed to remain free of the dead air region created by the horizontal stabilizer.

One point the report makes is that many pilots do not realize that there must be sufficient fixed area beneath the horizontal tail to dampen the spinning motion. This fixed area is necessary to "steepen and slow the equilibrium spin." As we know, planes are easier to recover from steep spins as opposed to flatter spins. Rudder is then used to change the moment, which initiates recovery. The report also points out that as the center of gravity moves toward the rear of the plane, it will spin flatter. As the spin transitions to flat modes, the elevator's effectiveness is typically reduced. In fact, in some cases the elevators become completely ineffective in a flat spin.

Studies reviewed by the report's author documented that the simultaneous application of counter-spin rudder and forward elevator control produced unwanted recovery characteristics. The report also found that occasionally counter-spin rudder application, while maintaining up elevator, gave satisfactory recovery results. The reason for these differing spin recovery characteristics may be that the rudder was shielded as a result of the downward movement of the elevator when it was applied at the same time rudder was applied. This is the reason that many aircraft spin recovery techniques require that counter-spin rudder is applied first, followed later by forward elevator control input. The report recommends that this interval be approximately one-half turn of the spin.

You can begin to see how important it is to know the correct recovery procedures for the plane you fly. One popular trainer, while certified for spins, has proven to be somewhat difficult in spin recovery. One NTSB accident investigation report related to this trainer stated that the use of standard spin recovery techniques are not effective in spin recovery. In those cases in which the standard recovery procedures were used, it was necessary to place the controls back in the pro-spin configuration, then re-apply the manufacturer's recommended spin recovery procedure. Even when correctly applied, spin recovery in this airplane can take a significant amount of time. The blanketing of airflow over the rudder and elevators can make spin recovery more difficult if the tail surface has not been properly designed for spin recovery. This can necessitate the use of other spin recovery techniques to overcome those deficiencies.

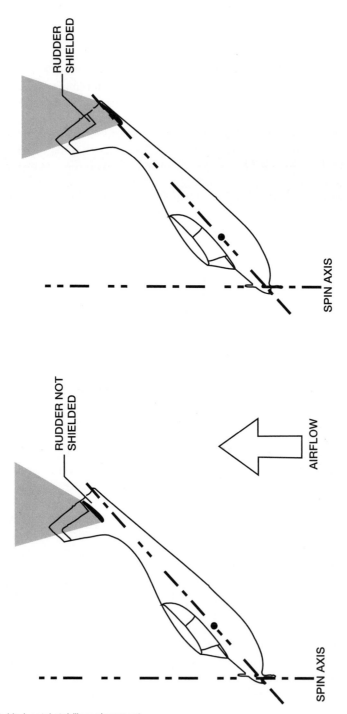

Fig. 8-2 Horizontal stabilizer placement.

The NASA report continues with a more in-depth review of tail design considerations. Rudder configuration is the first area of review. Two rudder installations are in use for most general aviation aircraft, the full-length rudder and partial-length rudders. As you might guess, full-length rudders extend to the bottom of the fuselage, while partial-length rudders "terminate at or above the top of the fuselage." When a plane is equipped with a full-length rudder, the area below the horizontal tail provides spin recovery capability. To improve the effectiveness of full-length rudders, horizontal tails should be placed high and forward in relation to the rudder. When a plane is equipped with a partial-length rudder the upper portion of the rudder "provides the most unshielded area." On aircraft with these types of rudder installations, the horizontal tail should be placed "low and rearward," resulting in the least shielding of the upper portion of the rudder by the horizontal tail.

The report makes an interesting point of the fact that when a plane is designed with "a low horizontal tail and a sweptback vertical tail, it is possible that almost the entire vertical tail, including the rudder, might be in the stalled wake of the horizontal tail." As you look over the flight line during your next visit to the airport, you will see this design can be found on many general aviation aircraft. The report notes that this does not make spin recovery impossible, due to the fact that the elevator "would still have some effect, particularly in the incipient spin. But the certainty of recovery would be jeopardized. . . . "

As noted earlier, you do not want to spin aircraft that are not certified for spins. Also, manufacturers are required to demonstrate only that their planes will recover from a one-turn or 3-second spin, whichever is longer. This report sheds additional light on what can happen when a pilot ignores the spin restriction for planes. "For airplanes with partial-length rudders and often for full-length rudders with a low horizontal tail, the rudder is usually mostly shielded by the horizontal tail and is, consequently, ineffective for spin recovery. Therefore, the elevator is relied on for most of the spin recovery."

Now comes a very interesting point. "Even so, an almost universal control technique suggested for recovery is rudder reversal followed by deflection of the elevators to neutral or down." This technique has proven effective when the recovery is attempted "before one turn is completed." "However, in many cases, it would be disastrous for the airplane to inadvertently wind up more than one turn because this technique may not recover the airplane if the spin has developed to two or more turns."

This particular point needs to be *re*-reinforced. You or other pilots you know may go out and spin airplanes that are not certified for spins; and because you begin recovery as soon as the plane enters the spin, you are able to get away with it. But the rudder may have little, if any, effect as part of the spin recovery, and it is only the fact that the elevators are effective that results in recovery from the spin. Once the spin becomes developed, the elevators can lose their effectiveness, essentially robbing the pilot of any control authority necessary for spin recovery. So if you attempt to spin a plane not certified for spins, you may find yourself in for a spin all the way to the ground because no matter what you do, it will not recover. This is a very blunt statement; but given the fact that there were ninety-six known spin-related accidents on file with the NTSB between 1/1/93 and 1/1/95, pilots need to begin to understand spins in order to help reduce the frequency of these accidents.

Testing has also allowed investigators to learn that the attitude of the plane during the spin can affect the "effectiveness of the elevator for spin recovery. In several documented spin test programs, good and rapid recoveries were obtained by rudder reversal and down elevator from spins that were steep and typical of median or forward center-of-gravity positions." These tests found that when a plane had a more rearward center of gravity, the spin becomes flatter and recovery from the spin was difficult or impossible. These tests also showed how the correct tail design is necessary for a plane to exhibit proper spin recovery characteristics. The report specifically mentions two tail modifications that were tested for spin recovery characteristics. Antispin fillets and ventral and dorsal fins were, in many cases, found to cause the plane to spin steeper, increasing the ability of the plane to recover from the spin.

Antispin fillets, shown in Figure 8-3, result in an increase in the damping of the tail. This, in turn, decreases the rate of the spin, resulting in a steeper spin. Planes normally recover more easily from steeper spins than flatter ones. The report also notes that the antispin fillets are most helpful when a plane has sufficient tail damping to begin with. On those aircraft where the tail damping power was well below that necessary for spin recovery, antispin fillets were not found to be useful. Ventral and dorsal fins had results similar to antispin fillets according to the report. If a tail was borderline on damping power, ventral and dorsal fins, as illustrated in Figure 8-4, would help slow the spin rate and increase the steepness of the spin. For those that were well below the required damping threshold, ventral and dorsal fins were not effective.

The report also looked at the effect of external wing tanks on an aircraft's spin and recovery characteristics. Two factors were considered, the aerodynamic effect and the effect as a result of the additional mass. The aerodynamic effect was found to be small, except in those cases in which the tanks had a negative impact to the plane's stability. In those cases in which stability was reduced, an airplane would occasionally be more prone to enter a spin. Mass, on the other hand, was found to have a greater effect on a plane's spin characteristics. This became more pronounced when the tanks were wing tip tanks. As you have seen earlier in this section, as a plane's mass distribution becomes more wing heavy, the spin recovery techniques change when compared to a zero loading aircraft. Given this situation, a plane with empty wing tanks may require that the standard counter-spin rudder, followed by down elevator, be used for spin recovery.

This same airplane may move into the wing-heavy category as the wing tanks are filled and require that the recovery technique of elevators down plus rudder against the direction of the spin be employed. Clearly, the pilot of this

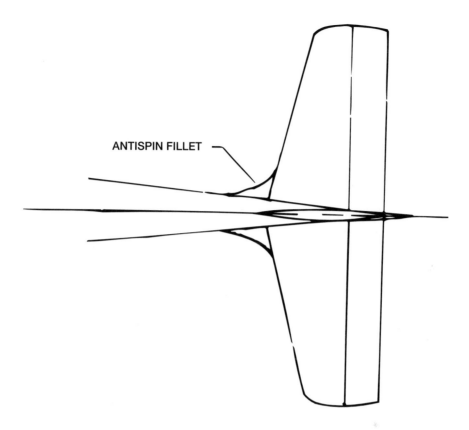

ANTISPIN FILLET

Fig. 8-3 Antispin fillets.

type of plane must know how fuel loading affects the plane's spin recovery characteristics. Until enough fuel burns out of the wing tanks, one set of spin recovery procedures might be necessary; while after the fuel load becomes lighter, another would be implemented.

Flaps can also affect how an aircraft spins. This NASA report supports the statement made in several previous chapters that when flaps are extended, a plane will usually spin in a flatter mode than when they are retracted. When the flaps are extended, the plane will also have a tendency to spin at a slower rate as well. The report points out that when the flaps are extended, the rudder may be less effective due to a larger wake behind the wing, decreasing the amount of air flowing over the rudder. Note that the report was discussing results from low-wing aircraft and these do not necessarily apply to high-wing aircraft.

One other point was made related to retractable gear aircraft. "Extension of the landing gear usually has little effect on the spin and recovery characteristics, but slight adverse effects have been seen from lowering the landing gear

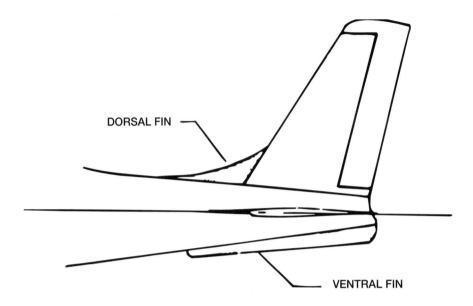

DORSAL FIN

VENTRAL FIN

Fig. 8-4 Ventral and dorsal fins.

on some airplanes. Therefore, it is generally recommended that the gear be kept in the retracted position when possible." At the beginning of Chapter 7, on flat spins, I touched on the myth that when a plane is in a flat spin, lowering the gear might help get it out. According to the NASA report, this is not typically true and may serve to make the situation worse.

Concerning the placement of the wing and tail empennage, the report found that high-wing airplanes tended to have better spin and recovery characteristics than low-wing planes. This was attributed to the fact that the "wake from a high wing is believed to pass above the tail so that the tail surfaces are not appreciably affected and the rudder and elevators are more effective in the spin recovery." It was also found that tail length can affect a plane's spin recovery characteristics. The longer length of the tail tends to cause the plane to spin at lower angles of attack and a higher rate of rotation. Shorter tails "cause the airplane to spin flatter and at a slower rate of rotation."

There are two final topics to review from the NASA report. The first is the position of the center of gravity. The more aft is the center of gravity, the greater the tendency for the spin to enter a flat mode. The report supports this standpoint, indicating that as a plane spins flatter, the effectiveness of the controls and their ability to recover the plane from the spin are reduced. Finally, the report notes that, dependent on the relationship between the location of the thrust line and center of gravity, use of engine power could have favorable or adverse effects on a spin. Given the unpredictability of the use of engine power on spins, it "is generally recommended that for single-engine airplanes, the throttle be retarded to the idle position during a spin."

SUMMARY

This chapter is less about "how to" and more about "why a plane behaves certain ways in a spin." A number of different stall/spin accidents were reviewed from the files of the NTSB. Each of these was included to point out a different aspect of what can cause spin accidents. Weight and balance outside of the center-of-gravity envelope, pilot misuse of controls, and inattention to the plane have all contributed to documented stall/spin accidents. Many of the accidents could have been prevented if the pilot had taken the proper steps or taken steps earlier.

Each accident review pointed out some of the errors that the pilot made, in addition to tying the example to points made in previous chapters. You must understand how the plane you fly reacts during various flight modes. Slow flight, stalls, and spins are areas pilots need to understand and be familiar with. Lack of knowledge about the plane's "feel," or reaction of a plane during a given flight situation, can put the plane and pilot into an unwanted accident.

The second portion of the chapter discusses the findings of a NASA report that dealt with the spin characteristics of general aviation aircraft. Oddly enough, the report pointed out that few studies on this topic have been completed in recent years. Much of the information contained in the report was collected from studies undertaken during the 1940s. However, the data still applies to today's aircraft.

Among the points made in earlier chapters that are supported by this report are the correct use of controls for spin recovery. Opposite rudder followed by down elevator was recommended as the most successful method for spin recovery on most zero-loading aircraft. Correct spin recovery procedures can change, based on whether a plane is fuselage heavy or wing heavy. Most single-engine, general aviation aircraft are zero loading, though, so the procedures discussed in the book typically hold true. Be sure to review your airplane's recommended spin recovery procedures before attempting spins in it. Certain tail designs might require different spin recovery techniques. If that is true for the plane you fly, be sure to use the manufacturer's recommended techniques.

Elevators lose their effectiveness the flatter a spin becomes, and rudder design can affect the amount of power it will have during spin recovery. During a spin, the elevator and wings can blanket the air flowing over the rudder; therefore, proper placement of the elevator can increase its effectiveness. Other factors, such as center of gravity, wing flap position, and extension of landing gear, are also covered. Rearward center of gravity and extended flaps can cause an airplane's attitude to flatten during a spin. If the design of the tail surfaces does not provide sufficient power to overcome the spin's rotation, it will not be able to recover from the flat spin.

The next time you practice spins, apply the information from this chapter to help you understand why the plane is behaving how it is. This text cannot begin to address all of the aerodynamic factors affecting a plane during a spin, but it attempts to support a basic understanding of the aerodynamics related to spins.

9

Successful Spin Training Factors

I FIND SPIN TRAINING TO BE FUN AND TRY TO MAKE IT THAT WAY FOR my students. Pilots have many reasons for wanting to become proficient at spin execution: curiosity concerning spins, a desire to become a safer pilot, interest in unusual attitude training, or concern over having accidentally entered a spin or near spin.

This chapter reviews some of these reasons, plus factors you should consider as you look for a flight school to give you spin training. The decision-making process as related to flying is also covered. The previous chapter reviewed seven different accident situations and discussed how many, if not all, of the accidents would have been avoidable if the pilot had changed the situation earlier in the chain of events leading up to the accident. The FAA has produced a very good publication related to the decision-making process. To help you better understand how this process takes place, and how it applies to stall/spin avoidance, some of this information is reviewed here.

REASONS FOR SPIN TRAINING

There are probably as many reasons for getting spin training as there are pilots. The next few sections examine some of these reasons, and then discuss what you should look for in spin training. The type of aircraft used, the flight instructor, and the curriculum can all affect the success of your spin training experience. In order to get the type of training you want, you need to communicate your requirements to the flight school, and determine if they are able to help you get the training you are requesting.

One of the most common reasons for getting spin training is a safety issue. Pilots may be looking for a better understanding of spins, what causes them, what they feel like, and what procedures are necessary to recover from them. They may have heard other pilots talking about them, read a book about spins, or rode along as someone else executed a spin. These are all great reasons for getting training. Throughout this book, experience in stall and spin entry and recovery has been established as one of the best ways to learn stall/spin avoidance. Chapter 8 emphasizes that all too often spin accidents happen while flying in the pattern, or when the airplane is at low altitude and slow airspeeds. But accidents also occur when the plane is at relatively high altitudes, in some cases several thousand feet. In both high- and low-altitude situations, spin recognition and avoidance can prevent accidents. Becoming familiar with spin entry and recovery can help you avoid these dangerous situations.

Unusual attitude training can also incorporate spin training into the curriculum. A number of pilots I have flown with have expressed a desire to experience what it is like to be in very unusual attitudes, from vertical up and down lines to inverted flight and spins. While concerned with safety, these individuals are interested in taking the training one step further. Not only do they want to learn how to deal with spins, they want to understand how to control an airplane safely in attitudes they may never experience outside of that training. Once again, though, having this type of training can be invaluable when attempting to control an airplane that is well outside of the straight-and-level flight regime.

To help students learn how to deal with these less-than-optimum situations, I will "mess up" aerobatic maneuvers the student may be practicing by putting the plane into an unexpected snap roll or pulling the engine power. This helps the student learn to react calmly and safely when faced with an unexpected change in attitude or power loss. Occasionally, these unexpected actions will put the plane into a spin of one sort or another, allowing them to learn to recover from spins when faced with one they had not planned on entering.

Some pilots want to learn spins to meet the requirements of a rating. Flight instructor candidates must log training in spin entry and recovery before they take their CFI checkride. The FAA requires that they learn only "spins," but some flight schools offer more advanced training to those candidates who are interested in becoming proficient in more than just a one-turn spin. The accidents reviewed in the last chapter document that even with flight instructors on board the aircraft, planes still get into spin-related accidents. The desire to get more complete spin training can motivate flight instructors to take additional training. While most CFI students I have taught are not overly concerned with learning inverted or flat spins, many take the opportunity to learn accelerated spins. Because it is possible that students will place these future CFIs into these situations, they find it beneficial to understand some of these more advanced spins.

Yet other pilots want spin training in the aircraft they regularly rent or own. If it is certified for spins, pilots should be able to find qualified flight instructors that are more than happy to give training in those aircraft. This gives them a feel for how the plane they fly on a regular basis will react during stalls and spins. The training may be limited to normal spins depending on the aircraft, but it will aid a pilot become proficient in spin avoidance in not only that plane, but others. Once you have learned the basics of spin entry and recovery, if you want to learn more exotic spins, you can continue spin training in more advanced aircraft.

Yet another motivation is the fun you can have. Oddly enough, enjoyment is why I first learned how to do spins. There is nothing better than stalling a plane, putting it into a spin, and watching the world turn around you. I can still remember the first spin I did in a J-3 Cub, and I knew then that the bug had bitten me hard. That was such a great little plane to spin, it was hard to believe not everyone who flew wanted to execute spins. Since then I have been fortunate enough to fly a number of different aircraft that were capable of executing spins. I learned how to spin each of them because it helped me learn how the plane handled, and it was a great way to entertain myself. Becoming proficient at spins will make you more relaxed with the plane, with what it is capable of, and increase your enjoyment when you fly.

You may have yet other reasons for wanting to learn spins. Whatever the reason, take the time to do it. You will be safer and more confident in your ability to fly a plane in unusual situations. Spin training will also help you have a better feel for how planes react when flying at slow airspeeds. You'll begin to learn how to fly an airplane in a coordinated manner, learn how to use rudder properly, and have a much better chance of avoiding an accident.

SPIN TRAINING

There are a number of different factors that go into your getting the type of spin training you want. The flight school, the aircraft the training is done in, the knowledge and experience level of the flight instructor, and the spin training curriculum all affect how successful your experience will be. While not everyone will learn to enjoy spins and view them as a source of entertainment, if you can't come away from your training feeling comfortable with spins and confident that you understand entry and recovery, your goals were probably not reached.

The Flight School

As we all know, flight schools run the gamut from one-man operations operating out of a hanger to large schools that train hundreds of students a year. These extremes, and every size in between, can offer high-quality flight training and professional instruction. So how do you determine if the school you are looking at is a good one?

One of the best ways is to talk to students and former students to see how they feel about the training they received. Find out if the curriculum they followed allowed them to successfully complete the rating for which they were training. You should also find out if the ground school portion of the training received as much attention as the flight training.

These are characteristics that separate good flight schools from poor ones. If a large percentage of students end up retaking written and flight exams, they are probably not getting the level of ground or flight instruction necessary from a flight school. This can be the result of insufficient information being passed on to the student. If a flight school uses flight time to give ground instruction, they are not properly focused on ground school. This results in the student wasting time and money trying to learn various pieces of information. Preflight discussions prepare a student for what will be done during the flight portion of the lesson. During preflight discussions, the instructor should clearly explain the objectives of the lesson and what the student's actions will be while flying. Critiquing should take place during and after the lesson, reviewing not only what the student needs to correct but what was done correctly.

You might also contact your local FAA office to learn more about a flight school. FAA personnel can give you information related to their experience with the school. While their involvement may not relate to direct experience in flight training, they will know if the school has been cited for major violations or other safety-related incidents. One poor report does not mean a flight school is not safety oriented; but if a pattern exists, you may want to dig deeper to find out if the school is lax in its safety standards.

You might find it beneficial to talk to the owner(s), manager(s), flight instructor(s), and other staff of the flight school. See if they have a positive attitude toward flight training, aircraft maintenance, and flight safety. If you find that they are cavalier in their approach to flying, you might want to consider how that fits in with your personal beliefs. Flight training should be fun and enjoyable; but if you find that the attitudes hearken back to the days of "anything goes" barn storming, this may not be the right atmosphere for accomplishing your goals.

Finally, find out if they have experience offering spin training and how comfortable they are giving this training. If you sense a reluctance on the part of the school instructors to offer it, they may not have the experience or comfort factor to be able to provide the type of training you want. The school should have a record of offering spin training in a safe manner. Ask the school how long and what types of spin training are available, in addition to the type of aircraft that are used for spins and how much experience the flight instructors have in spins. These items all factor in to how successful the flight school will be at providing quality spin training.

Training Aircraft

Depending on the type of spin training you want, the type of airplane used for training can vary a great deal. If you are interested in normal, upright spin training, potential training aircraft include the Cessna 152 or a Citabria. For more advanced training that includes inverted or flat spins, you will need to fly planes like the Pitts Special. As you review a flight school, look at the aircraft used for spin training and determine if the plane is certified for spins.

If the flight school tries to persuade you to learn spins in a plane not certified for spins, head in the opposite direction. You should also avoid advanced spin training in a plane certified for only normal spins. Chapter 8 explains that spin recovery is in part dependent on the design of the tail. A plane that is able to execute normal spins safely may not be able to execute flat spins and recover. Verify that the plane is designed for the type of spin training you want before you fly it. If you are uncertain, contact the manufacturer and ask them what types of spins it is certified for.

Once you have found a school with the proper types of aircraft, take a close look at the plane. Find out if the plane is being maintained well. You can do this by performing a thorough preflight of the airplane. As you walk around the plane, look for wrinkled skin or fabric, which might indicate structural damage to the plane. Look for popped or loose rivets, another indicator that the plane may have been subjected to abnormal load factors. Take a close look at the engine mount, looking for corrosion, worn rubber mounts, or bent tubing. Each of these could turn into a serious problem when a plane is subjected to spins. Check the controls for proper movement through their full range of motion. Spin recovery is not the time to find out that the elevator sticks when you pull it to the fully aft position. This might seem like a minor point, but there are aircraft that have demonstrated just that tendency, making spin recovery more exciting than it needs to be. Unless you move the controls through their entire range of motion during the preflight, you may not find this type of problem until you are actually in a spin.

Look over the plane's log books as well. These will tell you of any damage history, repairs to the plane, and if the plane's annual or 100-hour inspections are up to date. Many flight schools keep their planes' log books in the maintenance hanger or office, making it necessary for you to ask to see them. This is not only your right, but your responsibility as the pilot in command. Again, you want to look through the logs to find out if there are any problems with the plane that might make it unsafe to spin.

I recently found a plane that had the weight and balance incorrectly calculated. The wrong arm length had been used for equipment that had been added in the past, resulting in an inaccurate center-of-gravity calculation. While in this plane's case the actual center of gravity was not outside the acceptable envelope, having a plane outside the center-of-gravity range when it is subjected to spins is dangerous. In general, make sure the plane is properly certified and maintained. If you have questions about a particular plane, be sure to ask the flight school for the answers.

The Flight Instructor

Flight instructors are the primary means through which you will learn how to execute a particular maneuver, such as spins. Flight instructors range in their personality types, their ability to communicate, their understanding of the subject matter, and their level of experience. When you get flight training of any type, find a flight instructor with whom you feel comfortable. There is nothing worse than trying to get motivated to go fly because you are not compatible with your flight instructor. I have had students tell me, "Just go ahead and yell at me if I do something wrong. I've got no problem with being yelled at and I've had plenty of flight instructors yell at me." I don't believe in yelling at students; it is not conducive to learning and puts more stress on them. If you do not like flight instructors to yell, then tell them that. If they continue to do so, find another flight instructor. On the other hand, if you like to be yelled at, then find yourself an instructor that fits this mold.

Like flight schools, flight instructors vary in every aspect, such as how they present themselves, how they communicate, and how prepared they are for lessons. There are certain attributes you should look for in a good flight instructor, though. First is professionalism. Does he or she treat flight instruction as the responsibility it is, or is it just a way to build more hours and move on to the next flying job? If your flight instructor does not seem interested in your success, is continually unprepared for your lessons, or seems less than enthusiastic toward flight instructing, that person may not have the right attitude to get the job done for you.

Second is the instructor's ability to communicate. Your flight instructor should take the time before and after a flight to discuss what you are going to do during the flight and how well you did it. Flight instructors should be able to clearly present their ideas and provide answers to questions you might have. And if they don't know the answer to a question, they need to tell you that, and then find the answer. If your flight instructor attempts to cover the fact that a question cannot be answered, there is a problem. During the preflight discussion, the instructor should help you understand what the correct amount of control inputs are for a given maneuver, what you can expect to experience, and how the plane will react. This gives you some idea prior to the flight of what you will be expected to do. This communication should continue while you are flying. Teaching techniques will vary with each instructor, but they should all talk to you during the flight, critiquing what you have done correctly and what needs improvement. After the flight, any final critiquing or questions that need to be answered should be discussed. A good flight instructor will schedule flights to allow for these pre- and postflight discussions.

The third area related to flight instructors is their experience level. When you begin to get into advanced maneuvers, like spins, the instructor needs to have a strong understanding of the subject matter. Simply taking you up and letting you do a one-turn spin does not qualify as true spin training. While you may learn the mechanical movements for entering and recovering from spins, you have not really learned about spins. Ask questions about what causes spins, what can affect how a plane will spin, and what can happen as the result of various control inputs. If you get glossed-over answers, your instructor may not have the background to teach you what you want to learn. No instructor can answer every question you might ask; but when an instructor does not know the answer to a question, he or she should tell you that, and then try and find the answer to your question.

A good flight instructor will have you flying the vast majority of the time you are in the air. If you find that an instructor is spending more time impressing you with how well he/she flies, as opposed to having you at the controls, consider working with another instructor. Flight time should focus on you gaining experience executing maneuvers, with the instructor flying only a minimum of the time.

Spin Training Curriculum

When you decide to get spin training, you first need to decide what type of spin training you want. Spin training can take many different forms, from basic upright spin entry and recovery to all out, rock-and-roll, inverted spins or flat spins. Your final decision will be based on the motivation you have for wanting the training. Once you decide what your goals are, find a flight school that can help you meet them. As you talk to the school staff, ask them if they have a curriculum for spin training. While the course itself does not necessarily need to be written down, they should be able to explain to you what they are going to cover in detail. You should also let the flight school know what your spin training goals are and be sure that what they can teach you is compatible with what you want to learn.

The curriculum discussions should also include ground school and any training manuals that might be part of the course. Preflight ground school discussions should be part of every lesson and manuals can offer additional information that may help students visualize certain spin-related concepts. The average number of lessons that make up a spin training course will depend on what type of training is being offered. A spin introduction course might consist of ground school and an hour of flight time, giving the student an opportunity to experience spin entry and recovery. A more in-depth course might involve two or three hours of flight time, giving the student more exposure to not only spin entry and recovery, but possibly the more exotic spins discussed in this book.

In a three-hour time period, it may not be possible for you to become proficient in all of the spins, but you could be exposed to each of them. You might need up to five hours or more of flight time to become comfortable with each of the different spins covered in the book. If your goal is to reach this level of proficiency, you need to invest a certain amount of time to accomplish it.

Part of the learning experience should also include spin avoidance training. You should be able to recognize when the plane is approaching a stall/spin situation. Aircraft sounds and flight control effectiveness as the plane slows are only two of the factors that can alert you to an impending stall. Proper use of rudder to control any wing drop off to prevent a spin is also important. The right curriculum offered by the right flight school can help you learn a great deal about not only how to be comfortable with spins, but how to avoid them.

PILOT JUDGMENT

This section looks at some of the factors that affect pilot judgment during the decision-making process. As you read this section, keep in mind how these factors can affect your ability to react correctly to an unplanned stall/spin situation, especially while in the pattern during takeoff or landing. In general, as you read through the following material, you will see that accident situations are often preceded by a series of events or decisions by the pilot that lead up to an accident. In many cases, accidents could have been avoided if the chain of events had been changed early on. Studies have found that accidents are easier to avoid the earlier the chain of events has been interrupted, and are more difficult to avoid the further down the event chain a situation moves.

The report used as the basis for this section is part of the FAA Accident Prevention Program. Titled *Introduction to Pilot Judgment*, this document is available from many FAA offices and may be on hand at your local fixed base operator. To begin, the document defines good pilot judgment as:

> Pilot judgment is the process of recognizing and analyzing all available information about oneself, the aircraft, and the flying environment followed by the rational evaluation of alternatives to implement a timely decision which maximizes safety. Pilot judgment thus involves one's attitude toward risk taking and one's ability to evaluate risks and make decisions based upon one's knowledge, skills and experience. A judgment decision always involves a problem or a choice, and an unknown element, usually a time constraint, and stress. (*Introduction to Pilot Judgment*, p. 2)

In short, pilots must continually assess themselves, the plane, and the environment that relates to the flight before, during, and after a flight and make decisions based on this information that result in the safe outcome of the flight. The environment can include any number of factors, such as weather, air traffic control, or passengers. Everything besides the aircraft is considered part of the environment.

Each of these areas, the pilot, the airplane, and the environment, are factors in pilot judgment. As already stated in this chapter, accidents are frequently the outcome of a series of poor judgments. This poor judgment behavior (PJ) chain can often be traced back from the accident to a starting point. The following example was cited in the report:

> A noninstrument rated pilot, with limited experience flying in adverse weather, wants to arrive at his destination by a certain time, and he is already 30 minutes late. In spite of his weather inexperience, he decides to fly through an area of possible thunderstorms and will reach this area just before dark. Arriving in the thunderstorm area, he encounters lightning, turbulence, and heavy clouds. Night is approaching, and the thick cloud cover makes it very dark. In the darkness and turbulence, the pilot became spatially disoriented because he failed to trust his instruments. (*Introduction to Pilot Judgment*, p. 3)

This example shows that the pilot made a number of errors in judgment that led up to the accident. First, he let his schedule pressure him into making the flight into deteriorating weather conditions. At one time or another, "get-home-itis" has affected every pilot's decision to make a flight and this can begin the chain of events leading to an accident. In this example, even after the flight started, options remained that would have helped prevent the situation from deteriorating further, such as detouring around the weather, landing at another airport, or postponing the flight until the weather had passed. Instead, the pilot opted to continue to fly on through the area affected by potentially poor weather conditions. Get-home-itis is a stress factor that can be controlled by understanding that being late, or not flying at all, substantially increases the safety of a flight.

The second judgment error committed by the pilot was continuing on into the thunderstorms once he encountered them. At that point, he still had the option to turn back or detour around them, but once again he decided to continue on his original course into the thunderstorms. The series of poor judgments leading toward a potentially serious situation here is painfully obvious. The third major error in thinking was not trusting the plane's instruments once it was necessary to fly using them as reference. This last error led to disorientation on the part of the pilot. Anyone who has flown on instruments has experienced the feeling at one time or another that the instruments are not accurate. It requires discipline and concentration to overcome this, but eventually the feelings pass and the body and mind adjust to what the instruments are telling them.

You may be able to point out other errors the pilot made in this example. While this example was fictitious, in reality it has happened all too many times, often with fatal outcomes. The longer a PJ chain goes unbroken, the lower the chances become to avoid an accident. One of the most important steps a pilot can take in breaking the PJ chain is to recognize that he is setting the chain up, and then change the pattern by using good judgment in the decision process. These two principles are connected to the PJ chain:

> One poor judgment increases the probability that another will follow. Judgments are based on information the pilot has about himself, the aircraft, and the environment, and the pilot is less likely to make a poor judgment if this information is accurate. Thus, one poor judgment increases the availability of false information which may then negatively influence judgments that follow. (*Introduction to Pilot Judgment*, p. 3)
>
> As the PJ Chain grows, the alternatives for safe flight decrease. If a pilot selects only one alternative among several, the option to select the remaining alternatives may be lost. For example, if a pilot makes a poor judgment and flies into hazardous weather, the alternative to circumnavigate the weather is automatically lost. (*Introduction to Pilot Judgment*, p. 3)

These principles can be applied to stall/spin scenarios as well. If a pilot decides to load an airplane with an aft center of gravity and over gross weight, the option to recover the plane if it stalls on takeoff may very well be lost. A decision to spin an airplane that is not certified for spins may also put the pilot into a situation from which there are no options for recovery. Executing these maneuvers too low to the ground may also reduce recovery options. Related to stalls and spins, good judgment will help you avoid situations that place you, your passengers, and the plane at risk.

Studies have shown that there are five attitudes that can be hazardous when it comes to sound decision making. Each of these attitudes influences the pilot's judgment process in a negative manner and increases the probability that a poor decision will be made. These attitudes are (*Introduction to Pilot Judgment*, p. 10):

- Anti-authority: "Don't tell me!"
  This attitude is found in people who do not like anyone telling them what to do. They think, "Don't tell me!" In a sense, they are saying "No one call tell me what to do." The person who thinks, "Don't tell me," may either be resentful of having someone tell him or her what to do or may just regard rules, regulations, and procedures as silly or unnecessary. However, it is always your prerogative to question authority if you feel it is in error.
- Impulsivity: "Do something—quickly!"
  This is the attitude of people who frequently feel the need to do something, anything, immediately. They do not stop to think about what they are about to do; they do not select the best alternative. They do the first thing that comes to mind.
- Invulnerability: "It won't happen to me."
  Many people feel that accidents happen to others but never to them. They know accidents can happen, and they know that anyone can be affected; but they never really feel or believe that they will be involved. Pilots who think this way are more likely to take chances and run unwise risks.
- Macho: "I can do it."
  People who are always trying to prove that they are better than anyone else think they can prove themselves by taking risks and by trying to impress others. While this attitude is thought to be a male characteristic, women are equally susceptible.
- Resignation: "What's the use?"
  People who think, "What's the use?" do not see themselves as making a great deal of difference in what happens to them. When things go well, they think it is good luck. When things go badly, they attribute it to bad luck or feel that someone is "out to get them." They leave the action to others, for better or worse. Sometimes such individuals will even go along with unreasonable requests just to be a "nice guy."

You can probably think of how these hazardous attitudes might factor in to stall/spin accidents. The "Don't tell me!" attitude can cause a pilot to ignore the weight and balance limitations of a plane, loading it in such a manner that could cause the plane to stall on takeoff or be incapable of climbing out of ground effect. This same attitude might cause a pilot to not use coordinated control inputs because he/she does not feel like flying in that manner.

The "Do something—quickly" mindset can lead a pilot to react to a stall situation improperly, without understanding the situation that caused the stall or taking the correct recovery steps. The "It won't happen to me" attitude can lead to a variety of situations, such as ignoring the airspeed and flying too slowly during takeoff and landing. The macho "I can do it attitude" can result in the pilot pushing the limits of a plane to the point that it could stall or spin as a result of airspeed loss or improper use of the controls. Finally, "What's the use" can cause a pilot to quit flying the plane once a stall or spin takes place, believing that there is nothing he/she can do to salvage the situation.

To help avoid these dangerous attitudes, the FAA document (p.16) has antidotes that pilots should use instead of their impulse. These are:

Hazardous Attitude	Antidote
Don't tell me!	Follow the rules. They are usually right.
Do something—quickly!	Not so fast. Think first.
It won't happen to me.	It could happen to me.
I can do it.	Taking chances is foolish.
What the use?	I'm not helpless. I can make a difference.

The document recommends that you memorize the antidotes so that they will come to you automatically when you need them. If possible, you should attempt to obtain this document. Again, the title is *Introduction to Pilot Judgment*, FAA-P-8740-53. It contains a test that can help identify the hazardous attitudes to which an individual is prone.

The last subject to be covered from this FAA document is how you can identify and reduce stress, which can reduce a pilot's ability to analyze and react properly to a given situation. To summarize, there are three types of stress: physical, physiological, and psychological. Physical stress includes "conditions associated with the environment, such as temperature and humidity extremes, noise vibration, and lack of oxygen." Physiological stresses include "fatigue, lack of physical fitness, sleep loss, missed meals (leading to low blood sugar levels), and illness." Psychological stress is "related to social or emotional factors such as a death in the family, a divorce, a sick child, a demotion, etc. Also, they may be related to mental workload such as analyzing a problem, navigating an aircraft, or making decisions."(*Introduction to Pilot Judgment*, p. 17).

The effects of stress are cumulative on an individual. Every person reacts differently to stress and can deal with differing levels of stress. When the stress levels become too high, overload and panic set in. "Accidents often occur when flying task requirements exceed pilot capabilities" (p. 18). Symptoms of stress can surface in a number of different ways, but can include physical, emotional, or behavioral symptoms. There is no magic formula that can be used to increase the stress level an individual is able to deal with, but experience can help. Tasks you were given during your early pilot training may have seemed overwhelming at first, but they have become second nature as you gained additional experience. These facts suggest that every pilot should get proper stall and spin training, to help successfully deal with the stresses of stalls and spins. To close this section, the document included an anonymous quote I find insightful: "A superior pilot uses his superior judgment to avoid stressful situations which might call for the use of his superior skills" (p. 20). Keep this in mind next time you decide you don't mind taking a risk associated with a particular flight.

SUMMARY

This chapter covered two topics: factors associated with getting proper spin training, and attributes, both mental and physiological, that can contribute to stall/spin accidents. While seemingly widely separated topics, they are related in their bearing on pilots being able to recognize and avoid accidents. The first half of this chapter introduced factors you should consider as you look for a flight school from which to receive stall/spin training. The reputation of the flight school, the opinion of former students, the type of equipment used for training, the stall/spin training curriculum, and the flight instructor all must fit your criteria in order for the training to be successful. You should decide what your goals for spin training are, and communicate these goals to the flight school. While there will be a minimum that a flight school will want to teach you, if you are looking for additional training, they should be able to offer that to you as well. The amount of time necessary to complete training will vary based on an individual's background and learning capabilities, and what the pilot wants. Getting proper stall/spin training can not only increase the ability of a pilot to recognize and avoid inadvertent stalls and spins, but can also be an immense confidence builder.

The second half of the chapter dealt with factors that can result in an increased probability for an accident to take place. Five hazardous attitudes interfere with a good judgment process and the antidotes for these attitudes were noted. Stress can overload an individual's ability to make good decisions or to be able to cope at all with a given situation. Experience and training can help reduce stress levels, though, and increase a pilot's chances of making proper decisions during a given situation. This information was taken from an FAA publication, which you should obtain and review.

10

Spin Training Tips
for CFIs

THIS CHAPTER LOOKS AT SPIN TRAINING TIPS CFIS MIGHT FIND USEFUL as they teach spins. While this chapter is primarily intended for flight instructors, it can also be of benefit to the general reader. When you complete this chapter, you will understand the types of training techniques that work well in teaching spins. Among the topics covered are preflight discussions, in-flight training techniques, and postflight discussions. Other topics covered include the attributes related to spins that students have expressed the most difficulty in learning and what might be done to help them overcome these concerns. Curriculum development is also reviewed as related to spin training. This type of information can be useful even for those readers who are not flight instructors.

As you read through the chapter, consider how each of the points relates to spins. Even when teaching advanced spins, many of the same concepts and learning challenges exist for pilots. As subjects such as student disorientation during a spin or the need to keep initial lessons short due to physiological and psychological limitations are discussed, you will find that it is necessary to adapt lessons to each of the pilots you are teaching. Every student is different in their ability to absorb what takes place during spin training. Many pilots find that learning to execute spins properly can be more intense, both mentally and physically, than other types of flight training. To help students succeed in their goals, you need to be aware of their psychological and physiological state, and then adjust lessons to allow them to maximize their learning experience.

FLIGHT INSTRUCTOR EXPERIENCE

Like any good teacher, flight instructors must have a strong understanding of the subject matter they are teaching. Spin training can place more demand on the student, the plane, and the instructor than many other types of flight training. The instructor must not only be able to execute spins as though they were second nature, but be able to point out what is taking place at any given moment to the student during the course of a spin. In order to achieve this level of spin knowledge, awareness, and comfort, the flight instructor must have executed a sufficient number of spins, gaining enough experience so that all of this falls into place. The foundation for the instructor's own spin experience must have been laid with proper spin training. Misinformation passed from a flight instructor to a student because of a lack of understanding on the part of the instructor can cause long-term misconceptions on the part of the student. We all know it is easier to learn something new than to unlearn something wrong. What you teach students as they execute spins will influence them for the rest of their flying careers, so you must be certain about the subject matter.

The ability to execute spins in a plane must also be complemented by a strong knowledge of the subject matter. Like many pilots, my first introduction to spins was not the result of a structured spin training course, but took place flying with a friend. I executed spins on a regular basis after that, not realizing how little I actually knew about them. Later, in preparation for my CFI, I received the normal one hour of spin training and logbook signoff. Again, while this instruction met the regulations, I still had a great deal to learn.

Due to my interest in spins and aerobatics in general, I continued to read to increase my understanding about spins, researching the information that was available. This is what is often necessary to understand what is taking place during spins and what the results of your actions might be. No single book can cover all aspects of spins; but through a continued review of material that is available, you can build a solid understanding. The effort invested in learning this information can benefit both you and your students as you train them.

As you review material you will probably notice that there are differing opinions about various spin characteristics. Some of these opinions are based on an author's experience or conceptions. At some point, you need to weigh these views to find out what is most correct for you. After talking with NASA personnel about spin studies that agency has performed, I found that there are still unanswered questions. While some areas of spins are well documented, others require more testing. For that reason, it is best for you to review as many spin-related documents as possible to build the most complete picture possible about spins. This, coupled with solid experience behind the controls of an aircraft, will allow you to give students the level of instruction they require to become proficient at spin execution.

SPIN TRAINING CURRICULUM

This section reviews topics related to the spin training curriculum you might design as you plan a spin training course. These topics include what your goals are as you design a spin training curriculum and the use of spin training manuals. In any curriculum, you must decide what you want the results to be before you can build the component lessons of the course. With this in mind, you should consider several factors as you design the lesson plans. These might include your spin experience, the type of aircraft you have available to you, and the needs that students have for spin training.

If your spin experience has centered around upright spins, then you do not want to offer training that includes inverted, accelerated, or flat spins. However, if your background includes well-founded experience in the more exotic types of spins, then you may want to incorporate these into the lesson plans. If you have not received training in spins that you want to offer, get it and get plenty of experience executing those spins before you attempt to teach them. Even with hundreds of spins under your belt, students can still surprise you and put the plane into a situation you may not be familiar with. The more experience you have, the better you will be able to cope with these unusual situations. Do not attempt to teach yourself spins. Many factors affect the successful outcome of spins, particularly the more exotic versions. For this reason, you want a qualified flight instructor to work with you as you learn a spin to help you avoid the potential pitfalls associated with each of them.

Once you have determined what your spin capabilities as a flight instructor are, consider the goals of the course you are designing. First, you need to know the spin capabilities of the aircraft available to you. If you have a Cessna 152 certified for spins, you need to design a course that centers around that plane. With a Decathelon available as a training aircraft, you might want to offer inverted spin training in addition to upright spins. More advanced aerobatic aircraft expand the envelope of spins you may want to include in the course, from normal upright spins through flat spins. Do not exceed the capabilities of the plane in an attempt to teach spins for which it is not certified. Once you begin to offer spin training that places you in the aerobatic realm, such as inverted spins, you also need to use parachutes when you fly. Plan on having this specialized equipment available if your course includes spins that are more involved than the upright version.

Finally, you should consider what your students might want from a spin training course. The curriculum I have designed moves pilots from standard upright spins to those as advanced as they are interested in learning. But experience has shown that students are frequently interested in a tailored course that allows them more flexibility. In the interest of providing this type of service to students, I am also willing to accommodate their needs, as long as the spins they learn fall within their capabilities and produce a safe pilot. For instance, some pilots want to learn to execute and recover from upright spins with great proficiency, but only want to ride through one or two examples of the other spins covered in this book. Their reasons are that they do not intend to execute these spins, but want to at least have been exposed to them.

You will also need to consider what texts would be useful in supplementing your flight training. FAA documents are available with some information on spins, but this can be scattered through any number of texts or reports and may be difficult to pull together. There are also texts like this one that are dedicated to spin explanation. Again, you will likely have to invest some time in order to find out what is available. The text, or texts, you choose should support the overall curriculum you are developing and enhance the student's understanding of what is taking place during a stall/spin.

Some texts focus on theory and equations, while others (like this one) cover the how's and why's of spins in a more practical manner. You will need to determine for yourself what teaching approach you would like to take, and then choose the appropriate texts.

In general, if you have students read the information about a particular spin prior to the lesson, they will have had some time to think about the activity and have questions for you prior to the flight. This can help them be better prepared for the lesson and increase the speed with which they become proficient at spins.

PREFLIGHT BRIEFINGS

This section reviews some ideas related to preflight briefings for spin training. During the flight portion of a spin lesson, students are placed in a rapidly changing environment. The more thoroughly they understand what is going to take place during the lesson before they leave the ground, the better they will be able to interpret and assimilate what is happening when they are flying. As flight instructors, we must take the responsibility of helping students learn seriously. To accomplish this, we must invest as much effort in talking with students before and after the flight as we do during the time we are in the air with them. Later we will look at suggestions for topics of discussion during the initial spin training briefing, in addition to covering student actions, aircraft actions, and sensory inputs. You will find that these subject areas can be covered as part of any spin training preflight discussion.

Initial Spin Training Flight Briefing

When a pilot comes to you for spin training, you should find out what his or her reasons are for wanting to learn spins and his/her expectations of spin training. Before I fly with a pilot, I like to go over these two areas for several reasons. First, it can help me understand student's background and give me some insight into the student's skill and comfort level with spins. You will find that some pilots are completely at ease with spin training, while others are somewhat afraid. If you find that a pilot is uneasy about the training, the first preflight briefing can be used to help the novice understand what we are going to be doing and increase their level of comfort with the lessons to come.

Years ago, a friend of mine went to a flight school, interested in taking flying lessons. The instructor, explaining nothing to the individual, took him up on the first lesson and put the plane in a spin. The instructor explained, "If this scares you, you don't want to learn how to fly." Not surprisingly, my friend never learned how to fly and after years of trying, I have not been able to talk him into going up with me. The instructor he flew with exhibited the worst of what an instructor can be, and cost aviation a pilot. If a pilot wants spin training, the last thing any instructor wants to do is make the student afraid to execute the maneuver. Some pilots will smile the entire time they are doing spins, but others will break into a sweat every time they pull the power back in preparation for a spin. To help reduce the anxiety level of pilots, you can spend time on the ground talking with them about what they are going to experience during the first lesson. I cannot express the need to do this too strongly. It is our responsibility to work as much as we can with students to help them become as comfortable as possible with spins.

In addition to potentially increasing the comfort level of the pilots during the first lesson's preflight briefing, you can also find out what their goals are. As discussed in the section related to spin curriculums earlier in this chapter, every pilot has different goals for spin training. By talking with your students during the first preflight briefing, you may be able to adjust the course to accommodate their needs, or better set the students' expectations about what the training will involve. This can help avoid student dissatisfaction and increase their enjoyment of the learning experience.

The first lesson preflight briefing is also an excellent time to cover the spin training curriculum(s) the flight school offers. Students may not be aware of what types of spins a plane can do, or the extent of the spin curriculum that you offer. By covering the curriculum, you may help increase their interest level and expose them to information they may not have. A discussion of the spin training curriculum can also make the students aware of the time commitment that will be necessary to accomplish various types of spin training, in addition to helping them understand the costs associated with training. If you offer a one-hour spin training course, cost may not be a large factor to the student. However, if you offer more in-depth training that may require three to five hours of flight time, the student should be made aware of the costs involved.

Standard Preflight Discussion

Prior to any spin training flight, you should sit down with the student and discuss what will take place during the flight. Some areas that you might include in the briefing are student actions during the flight, aircraft actions during the flight, and the types of sensory inputs students can expect to encounter as they execute the spins you will be covering during that lesson. There may be other topics you want to include, such as airspace utilization in the practice area, ATC procedures, and basically anything that can affect the flight. But as they relate to spins, a discussion of the three topics mentioned can help your students be better prepared for what they will experience during flight.

Related to the area of student actions, you can cover their actions prior to the spin, during spin entry, during the spin, and, finally, spin recovery. Actions prior to the spin might include items such as use of carburetor heat, clearing turns, and slowing the aircraft to a stall. In some cases, talking students through the actions as you sit in the plane while still on the ground, prior to starting can help them learn the procedures more quickly. If you are teaching inverted spins, you will also need to show them how to roll to inverted prior to stalling the plane. Student actions during spin entry can include the stall itself and use of rudder, elevator, and ailerons to initiate a clean entry into the spin. Once again, showing students as they sit with you on the ramp can help them understand exactly what they will need to do during each phase of the spin.

Show students how to use the controls once the plane has entered a spin as well. In most cases you will want them to maintain neutral ailerons, forward or aft elevator input depending on the spin, and pro-spin rudder. But if you are teaching knife-edge or flat spins, students will need to understand what actions they should initiate once the plane has entered the spin. Finally, showing students how to recover from the spin with proper use of controls can also reduce the amount of time it will take them to learn once you are in the air. The discussion of student actions should also include emergency procedures as well as any other actions they might need to take specific to the training or aircraft. By showing and explaining what actions you expect students to perform while on the ground, you can help them learn more rapidly in the air.

The second topic of interest, aircraft actions during the spin, is also important to discuss with students. The motions a plane enters while in a spin can be overwhelming to new spin students. If you make them aware of what the plane will do in response to control inputs, you can help them begin to anticipate the plane's reactions. For instance, as the plane stalls and rudder is used to initiate the spin, let them know what the plane's nose will do, whether it drops sharply or smoothly falls off into the spin. Will the wings bank steeply or do they tend to remain level during spin entry? Does the plane pass through inverted on its way to stabilizing in the spin, or is the pitch angle during the entry more gentle? During recovery, does the plane react quickly to spin recover inputs, or does it take one or more turns before the plane recovers from the spin? As you can see, by giving students this type of information, they will not be second guessing what is taking place and wondering if these actions are normal. The more students know about what they should expect from the plane as it spins, the less likely it is they will be surprised by what actually takes place.

The last area of interest, sensory inputs during the spin, will help students prepare for how they will feel during the spin. Students often are not prepared for and do not understand what they will physically experience during a spin. The rapid yawing, the sight of the ground spinning up at them, and the g forces pressing them into the seat may be foreign and unsettling to pilots until they learn this is normal for spins. By explaining to them the physical feelings they will encounter, the sounds of the engine, the potential disorientation during the spin due to a lack of reference points, or the feeling of the controls during the spin, students will again spend less time wondering whether what they are feeling is normal or if they should be concerned. The need to take the time to cover these areas becomes even greater as you get into more advanced spins. Students can quickly become disoriented if they are not able to assimilate their environment. While it may still seem very foreign, having some prior knowledge of what they will physically encounter during the spin will help them to feel more at ease during early spins.

Discussing what students can expect to see as they execute a spin can help them pick out reference points and reduce the normal disorientation most pilots experience during early spin training. By preparing a pilot for the nose down attitude, the rate of rotation the plane will exhibit, and how to pick out reference points to use outside of the plane during the spin, the pilot will execute the spin better. The motion of the plane should also be discussed prior to the spin. The severe yawing of the plane may have a tendency to throw a pilot about somewhat, and knowing this is normal can prepare the pilot for these actions, reducing their discomfort with the first few spins. Especially when you get into inverted spins or flat spins, you will find that pilots are quickly overloaded with sensory inputs and tend to focus on one or two aspects of what is taking place. G-loadings, negative g's in particular, will be more severe during some of the more exotic spins. Prepare pilots for this by covering this in advance of the flight as well.

There may be other items you will want to cover in the preflight briefings. While you cannot explain everything that will happen during a spin to a student prior to a flight, taking the time to have a thorough preflight discussion can improve the learning curve as students enter spin training.

STUDENT STATE OF MIND

Many factors can affect a pilot's decision-making capabilities and performance. This section looks at some of the typical psychological factors that can affect students while they are receiving spin training. The last section discussed how explaining what is going to happen as clearly as possible during the preflight briefing can help reduce a student's apprehension. For many pilots, spins represent an unknown, with many of the myths about spins overshadowing what they are going to learn. This apprehension can cause stress that detracts from the student's ability to learn. As you work with pilots, be aware that this stress or apprehension may manifest itself in many ways.

During the preflight briefing or the flight portion of the lesson, you should monitor the students. If they appear distracted, do not respond to questions, or fidget a great deal, they may be overly nervous about spin training. Even though I have explained what will happen during a spin prior to a lesson, I have had students let out a very audible "Whoa" as the plane enters a spin because the nose down attitude during the spin is a shock to them. Students can break out into a cold sweat during spins or begin to become nauseous. As the instructor, you need to be aware of when a student begins to demonstrate this behavior. You have several courses of action you can take to prevent the student's emotional and physical state from deteriorating further.

The first step is to discuss what the student is feeling, in a tactful manner. I try to reassure students that the feelings they are having are normal, and that they are not the first to encounter them. Besides being true, it helps to bolster the students' confidence, which can begin to suffer if they are having problems becoming comfortable with spins. Before, during, or after a flight, it may be necessary to talk to students if they are feeling low about what took place during a lesson. If the student feels too badly, either physically or mentally, it may be necessary to cut the lesson short, and then review what went wrong and proceed from that point. In some cases having a confidence-building talk on the ground is better than continuing to push the student in the air, hoping he or she will begin to relax. Even if it does become necessary to cut a lesson short, be sure to keep building on what the student has learned and the positive points of the lesson. When giving any training, and especially spin training, build the students' good feelings about what they have accomplished.

If students begin to feel physical discomfort, normally related to their stomach, they can also react in several different ways. When I give spin training, I frequently ask, in a very positive manner, how the student is feeling. If students become quiet, or begin to slow in their reactions, they may be feeling physical discomfort. In many cases, students do not want to admit this to themselves or to you. When this happens, you can suggest that your student fly straight and level for a while, to see if things "settle down." Some students recover and are able to continue flying. Even when students do recover, they may have a tendency to fall prey to discomfort again fairly quickly. When a student does not seem motivated to continue to fly, it may be best to call it a day and head back to the barn. I have had students visibly relax once the call to return to the airport has been made.

When you are doing spin training with a student, during the preflight briefing you should consider letting the student know that initial lessons are normally shorter than other types of flight training. By making the length of the lesson dependent on how the student feels during the lesson, as opposed to the standard one-hour lesson length, you are removing the taboo that if students are unable to do spins for an hour, they are at fault. Especially during the first lesson, I tell students we will stay out at long as they feel good about what we are doing, whether that is one spin or an hour of spins.

I have had students call it a day after just two or three spins, but each time we flew they were able to increase their tolerance to spins. The whole point is to be flexible in how you train students for spins. If they need to fly straight and level to let the world settle down, let them do that. When a student gets quiet or doesn't respond to instructions, it may be time to pack it in for the day. At the other end of the spectrum are students who can fly spins for an hour straight and still want to keep going. Once again, adjust the lesson to help individual students feel as good as possible about what they are learning. Keep in mind that even when students feel physically up, their ability to learn can diminish after an hour.

SPIN DEMONSTRATIONS

This section reviews tips about how to teach students spins. The introduction of a new spin will normally begin with a demonstration of the particular spin. Even though I cover what will take place during the spin in the preflight briefing, I like to review the spin again during clearing turns to refresh the students' memory. I typically execute the first spin having the students keep their hands and feet on the controls as I fly the maneuver. As I pull the power back, I begin a running dialog indicating what I am doing with my hands and feet, and what I am looking at. This seems to help students focus on what is taking place, as opposed to wondering what they should be focusing on.

As the plane stalls, I continue to talk, telling them which rudder I am pushing and what I am doing with the control stick. During the spin, I also count off the position of the plane every half turn to help the students maintain their orientation. As I recover the plane from the spin, I once again tell the students what I am doing. To help keep the maneuver from overwhelming students, I keep initial spins to one or two rotations. This gives them an initial exposure to spins that is easier to absorb and is less overwhelming. It has also been my experience that when first exposed to a new spin, students like to have it demonstrated to them a second or third time before they feel they understand what is taking place enough to execute the spin themselves. If you have students who want to ride through the spin several times, this is a normal reaction.

In many cases, students will focus on one aspect of the spin at a time. During the first demonstration of a spin, students may concentrate so hard on what you are doing with the control stick, they do not pay attention to the rudder pedals, airspeed, or what is going on outside the plane. The second time they may notice the control stick and rudder actions, but still not be certain where the plane is during the spin. For this reason, it may be necessary to run through more than one spin example before individual students feels they have absorbed enough to execute it themselves. As you move each control surface during spin entry, while in the spin, and during recovery, tell your students what you are doing so they will notice what is taking place.

Once your students are ready to execute a spin, talking them through each control input that they should make during the spin may be helpful. Until novice pilots become more familiar with each of the steps required to enter and recover from a spin, they may be very tentative about making control inputs. Talking them through the steps until they are more confident can help them overcome uncertainty. Once the plane stalls and enters a spin, be sure to count off each half rotation to help the student maintain orientation during the spin. After the student recovers and you are climbing back to altitude for the next spin is a good time to critique the student's actions during the spin. In this manner, the spin is still fresh in their minds and they will be able to correlate what you are telling them to what took place during the spin.

In some cases it may be necessary to help students with control inputs as they fly the spin. I have found that, especially related to rudder use, students may make only partial control inputs. This can prevent the plane from properly entering a spin. If you find that helping a student make a control input is necessary, be sure to let the student know you are helping with that input. Your reminder can help make students more conscious of the amount of control input necessary to execute the maneuver properly. Keep in mind that letting them make mistakes can also help students learn what the plane does when the controls are not used correctly. You will need to decide which course of action is best, based on what has taken place with an individual student.

The instruction techniques covered so far work well for not only normal spin instruction, but also when teaching advanced spins. You will probably find that pilots become disoriented more rapidly when performing maneuvers such as inverted or flat spins. For this reason, you should be certain to talk to your students during the spin to help them absorb what is taking place while the plane spins. As the instructor, you will also need to be situationally aware of what the plane and student are doing while in the spin. Your altitude, the condition of the spin, and how the student is reacting must all be monitored as the spin progresses.

During their initial exposure to more advanced spins, some students get so far behind the airplane they do not remember what to do next. In these cases, remind them what needs to happen, and then if they do not respond in a reasonable period of time, help them out as you explain what you are doing. I must stress this happens only in a minority of teaching situations, but you are the ultimate judge and must keep the situation safe at all times. Every instructor has a hair-raising tale to tell about student training, but spin training requires that you stay focused at all times and know what your outs are if the situation does not work out quite right. Know the lowest altitude you will let the plane go to before recovery from a spin, and then stick to that.

POSTFLIGHT DISCUSSION

Once the lesson is over and you are flying back to the airport from the practice area is a good time to begin discussions about what took place during the lesson. The students' actions are fresh in their minds, and they are still focused on what took place as they flew. In addition to your offering your critique of the lesson, let the pilot give his/her impressions of what took place. This can give you some insight into how students perceive what they have learned, and how they performed. This student self-critiquing can help you determine what additional information needs to be covered to resolve any problems or misconceptions a pilot might have about what was just learned. Once you are on the ground and the engine is shut down, if the student requires any demonstration of control movements you can cover this before you exit the airplane. After the student pilot's questions have been answered, discuss what you plan to cover during the next lesson and suggest reading material the pilot may want to review prior to the lesson.

SUMMARY

This chapter discusses some areas of interest to flight instructors who are involved in spin training. As with the rest of the book, this is not intended to be a self-taught course in spins. Before you give spin training to others, you should have a very solid background in executing spins and a knowledge base that allows you to clearly explain what is taking place during spins. The chapter covers how to set up a spin training curriculum, some ideas for pre- and post-flight discussions with students, and how to teach spins to students. Covering every aspect of flight training as they apply to spins is impossible, but you should have some basic guidelines that can help you be more effective as a flight instructor.

Safety should always be foremost in your mind as you give spin training to students, and you should ingrain this in the mind of every pilot with whom you fly. As you help other pilots learn how to successfully execute spin entry and recovery, be sure to cover clearing turns and minimum altitudes necessary for safe spin execution. By providing a professional, complete spin training curriculum, you will be able to help pilots of all experience levels become safer, more confident airmen.

Conclusion

FLIGHT SAFETY SHOULD BE OUR MAJOR CONSIDERATION EACH TIME we fly an airplane. Whether we fly for business, pleasure, or sport, a common theme for every flight should be to ensure that we have done everything possible to safely begin and complete each flight. My instructors over the years drove that point home, and I have tried to accomplish the same task with each of my students. As I stated at the beginning of the book, I have been fortunate with the instructors I have had during flight training and I feel they laid a solid foundation of good habits and understanding.

I take the responsibility I have as a flight instructor very seriously, and this book has been an attempt to make others aware of the need for flight safety. Accidental stalls and spins result too often each year as a result of lack of training and knowledge. Pilots today are frequently given only cursory knowledge of spin entry and recovery. Stall training is given more consideration during flight training; but as a result of talking with students and flight instructors, I feel even stalls are not covered to the level of detail necessary. Training in stalls and recovery from them do not give students the level of understanding they need to be competent at these maneuvers.

This book is not intended to be a self-taught course in stalls or spins. Instead, it should be used in conjunction with a qualified flight instructor, in a properly certified airplane, to further your understanding and competency at stall and spin entry and recovery. As a result of this knowledge and flight experience, you will understand what a plane "feels like" as it approaches a stall or spin. The proper training and practice can help this feel for the airplane become second nature, giving you the ability to avoid a complete stall or spin. And proper reactions to an unplanned stall or spin can greatly increase your chances of safely recovering from them.

Hopefully, your knowledge related to stalls and spins was challenged in the course of reading this book. In many cases, the examples for entry and recovery from a stall or spin apply to the specific aircraft used in the examples given in the book. You will find as you further research the topic of stalls and spins that there is no one right answer for every stall/spin situation or airplane. What has been presented here can be used as guidelines, but never assume that you have learned all there is to know about stalls or spins. Over the years, and especially as a result of conversations with those who are experts in the field of spins, I have learned that I must continue to learn, because even the experts do not have all the answers.

I highly recommend that each of you get spin training from a qualified flight instructor. But I would like to suggest you take that one step further, and get training in basic aerobatics as well. Not only does spin training make you a safer pilot, but approached with the proper attitude, it is downright fun. In the very same manner, aerobatic training from a qualified flight instructor can help make you a safer pilot. Knowing how to react when the plane is in very unusual attitudes can be a great confidence builder. But the best reason to get aerobatic training is because it will put a permanent grin on your face. There is nothing like watching the earth and sky roll around you as you execute a loop, or watch the ground fall away as you execute the upline of a hammerhead. Aerobatics can give you a sense of what the airplane is doing that no other flying can.

Many of the spins covered in this book are aerobatic maneuvers. Part of the motivation for including them is to make pilots aware of some of the less well known spins. As you saw in the chapter that covered NTSB accident reports, pilots manage to put themselves into the more exotic spins, like the flat spin, without realizing how their actions will cause the plane to react. Again, each plane may react differently to the control inputs. Even the same plane may behave in a different manner depending on how it is loaded or a number of other factors. But realizing that the possibility of a plane entering these more exotic spins exists may help you avoid them.

Finally, remember that you are the pilot, and make the plane do what you want it to. Do not let the airplane fly you. In many accident situations, the pilot has been distracted from flying the airplane and loses control. Always keep the plane flying and attend to distractions as best you can. Flying is one of the most satisfying activities you can undertake, and you should take great pride in the professionalism you approach that activity with. Whether the flight you are about to begin is a hop around the patch or a cross-country flight, use the training and knowledge you have accumulated to make sound decisions. In the end, this will keep you flying for many years to come.

Bibliography

Airframe & Powerplant Mechanics Airframe Handbook. 1972. Federal Aviation Administration, Washington, D.C.

Airman's Information Manual. Revised 1994. Federal Aviation Administration, Washington, D.C.

Flight Training Handbook. Revised 1980. Federal Aviation Administration, Washington, D.C.

Gene Beggs. 1994. "Spinning with Gene Beggs." *Sport Aerobatics*, April 1994, pp. 31–36.

"Introduction to Pilot Judgment." *Accident Prevention Program.* Federal Aviation Administration, Washington D.C.

Pilot's Handbook of Aeronautical Knowledge. Revised 1980. Federal Aviation Administration, Washington, D.C.

Summary of Spin Technology as Related to Light General-Aviation Airplanes. NASA TN D-6575, December 1971, David S. Bowman Jr., Langley Research Center, Hampton, VA

Vince Page. 1994. "Letters," *Sport Aerobatics*, August 1994, p. 31

Index

S

T